The Cosmic Water Hole

The Cosmic Water Hole

Emmanuel Davoust

translated by Barbara Jachowicz

The MIT Press
Cambridge, Massachusetts
London, England

First MIT Press edition, 1991

Translated with a grant from Direction du Livre,
du Ministère Chargé de la Culture.

Originally published as *Silence au point d'eau* (Teknea, 1988). Set by Teknea from computer disks provided by the author and the translator. Printed and bound in the United States of America.

Library of Congress Cataloging-in-Publication Data

Davoust, Emmanuel
[Silence au point d'eau. English]
The cosmic water hole / Emmanuel Davoust – 1st MIT Press ed.
p. cm.
Translation of: Silence au point d'eau.
Includes bibliographical references and index.
ISBN 0-262-04114-6
1. Life on other planets. 2. Extraterrestrial anthropology. I. Title.
QB54.D3913 1991 90-42961
574.999-dc20 CIP

Contents

Foreword

Dream of a group of travelers in the desert, hungry for some sign that they are not alone. They find many oases, but each is strangely sterile. Even a foreign tongue would be welcome, but all that can be heard is the soft inanimate sighing of wind over the sands. No barking of dogs, no singing of birds, no calling of children, no hail of greeting — in fact, no life.

Wake from the dream. We are the travelers. The human race finds itself spiraling through an almost unimaginably vast desert. The universe is characterized by relentless extremes of heat and cold, high and low pressure, shattering velocities of encounter, blasting effects of super-energetic radiation — all seemingly hostile to any form of life. Yet there are, or at least there should be, some other cases such as the fragile planet we inhabit.

As of this writing no other solar system has been identified, but both theory and observation suggest that such systems should be ubiquitous. It follows that an almost countless number of planets would be dotted through the galaxies, and some of these would happen to be more or less like Earth. Given such potential sites, most modern biologists have no problem in visualizing at least a few of them spontaneously populating themselves with life forms feeding off the energy and entropy streams furnished by their central stars, developing ever-increasing complexity with the passing eons, and perhaps in rare cases achieving intelligence, even science and technology.

Such, for many decades, has been the conventional picture. And it may even be right. Since at least the 1920s, science fiction, some of high quality, has highlighted a few of the limitless possibilities. For the past 30 years increasingly serious scientific inquiry has been brought to bear on the question. In 1982, the prestigious International Astronomical Union established Commission 51 (now called the Exobiology Commission); almost immediately it became the largest of all the IAU commissions.

Not surprisingly, there are skeptics. For example, they point out that exobiology is "the only science without a subject matter." They question whether it ever will have a subject matter. Some critics contend that the actual possibility of life originating spontaneously is so indescribably small that it could happen, still with vast improbability, only once in our entire universe, and we of course are it.

Other dissenters, at the opposite extreme, accept a growing consensus on the probability of life appearing and evolving toward intelligence whenever conditions are favorable, but then point out that by this argument we should be Johnnys-come-lately. Millions if not even billions of years ago, civilizations should have arisen around other stars. By developing technology far beyond ours, they should long ago have populated the entire galaxy, either by themselves or with their surrogate robotic probes. But where are they? Does their apparent absence, and their silence (at least to our present ears), bespeak their nonexistence?

Still other critics stress the awesome distances between even nearby stars, the even more awesome amounts of energy and time required to send the smallest physical probes between them, and hence the possibility that quarantine is a universal rule for disparate life forms in the universe.

Why, then, are there exobiologists? For many reasons, probably foremost among them the importance of the question. *Are we alone?* This is surely one of the most profound inquiries humans can ever undertake. The negative is impossible to establish with absolute certainty. Nevertheless, if growing evidence over decades and centuries of search should fail to find traces of life elsewhere, this places an even greater responsibility on us and our descendants to keep the flame alive, to continue to offer the priceless gift of sentience to countless generations with potentialities perhaps beyond our imaginations. If the search should succeed, it will become arguably the most important study of the human race to learn the properties of totally alien life and to try to establish meaningful contact.

In addition, the field of exobiology has both targets and tools. In the solar system are several possible sites for at least primitive indigenous life forms, most notably the Martian subsurface and the warm oceans and possible volcanic vents beneath the relatively thin icy crust of Jupiter's satellite Europa. In addition, a shell of electromagnetic radiation — broadcasting of many kinds, continuing to expand at the speed of light — already announces our presence to any curious listeners around any of the thousands of stars currently lying within 50 or 60 light-years of the Sun. Conversely, should they be broadcasting, we could equally well tune in on them; hence the subfield of SETI, the Search for Extra-Terrestrial Intelligence.

The practical problems facing SETI practitioners are severe. Advanced civilizations communicating from planet to planet or from planet to spaceship will surely use, if only for reasons of economy, exquisitely sharp pencil beams of transmission which have vanishingly small probabilities of intersecting the Earth. Our hopes must presumably rest on our happening to lie in the relatively broader beams of radars or of true broadcasting, or even on the unknowable but surely trifingly small chance of the presence of a deliberately omnidirectional beacon. To compound the search problem, any of trillions of frequencies might be used by extraterrestrials, and their sources might lie in any of billions of very specific directions as seen by us. Their transmissions might be so unfamiliar or so subtly coded as to look essentially like random noise to us. Yet more troubling, considering all the billions of years which have gone before and which are yet to come, we have little reason to hope that the presumably rare occurrence of an intelligent radio-transmitting civilization reasonably near the Sun would oblige us by existing during just this the brief span of decades during which our interests and abilities have been aroused. Despite such problems, more than fifty SETI searches have already been mounted, with ever-increasing sophistication.

Within the last few years, relatively low-cost microchip technologies have offered the ability to expand the power of these searches by roughly a millionfold. More than a decade ago I was privileged to chair a distinguished subcommittee of the U.S. National Research Council to consider whether SETI had become a valid field for professional astronomical study and funding. After a year of investigation and consideration we concluded in essence that, while the probabilities of any successful detection are very low, we now have for the first time some realistic prospect of sounding the multidimensional cosmic haystack. Because of the overwhelming importance of any positive answer, we recommended that the field be modestly pursued. NASA agreed. However, at this point the greatest problem of all appeared, namely the ridicule of influential representatives. Senator William Proxmire presented his Golden Fleece Award to SETI, and NASA was forbidden to undertake investigations in this area. The problem has not gone away. Reading the transcript of a so-called debate in the House of Representatives (*Congressional Record*, June 18, 1990, p. H4329), which included extensive quotes from supermarket tabloids, was a sobering experience for my still-strong faith in representative democracy. Suffice it to say that as of 1990 a realistic SETI program has yet to be funded. But the dream lives on. Many are trying in various ways to bring it to some degree of reality. And, even if strong Earth-based searches are mounted and fail to find a trace, in our children's time there will surely be developed, in the deep radio quiet of the far side of the Moon, SETI observatories able to speak with great authority to this awesome question.

Meanwhile, a number of books have explored the subject. Few if any have dealt with it from so many different points of view as the present work. Readers will not learn whether there is life on other worlds, but they will come away with a deeper realization of the antiquity, the richness, the complexity, and the importance of the question. For now, we must be content with that.

Harlan J. Smith

Acknowledgments

The facts and ideas which make up the architecture of this book accumulated bit by bit, as a result of my reading and of my participation in conferences and debates. Feedback from the public helped me clarify my thoughts. Joachim Cardona carefully read several successive versions of the French manuscript; his questions and suggestions helped me improve both the form and the content, and I am very grateful to him. I would like to thank Dr. Philippa Gander, who read chapters 2, 3, and 4 and corrected my hasty and sometimes incorrect conclusions in biology; Hans-Ulrich Daniel, who contributed many constructive comments on the English text; my father, Eugene Davoust, who eliminated most of the stylistic and spelling mistakes in the French edition (I am responsible for the remaining ones); and my colleagues at Besançon Observatory, who contributed to realizing this book. NASA, Editions Belin, Lucienne Gouguenheim, and the French Astronomical Society supplied illustrations for the book free of charge, and I thank them for it. Last, but not least, I thank my wife, Barbara, who read and reread many versions of the manuscript, corrected numerous errors and awkwardnesses, translated the book into English, and accepted my sacrificing many evenings and weekends to writing this book.

Introduction

. . . we live with ideas that, if we really felt them, would disrupt our whole lives.

—Albert Camus, *The Myth of Sisyphus*

Are we alone in the universe? The question comes to mind naturally when we look up at a starry sky. And indeed, among those thousands of stars, why wouldn't there be a sun like ours, surrounded by planets, some of them inhabited by beings like ourselves? This question is among those that man has been asking since time immemorial about life, death, and the beyond, without ever leaving the realm of speculation. These questions reflect the need to put some order into the world in which we live, but also to question the framework of our everyday life. What if radio astronomers finally intercepted an intelligible message from the outskirts of our galaxy? Our view of the world would be completely upset.

This subject, which until now seemed to be reserved for philosophers and writers of science fiction, has been opening up to astronomers in the past 20 years. But where should we begin the search for life in the universe? A discovery or a new fact is often the starting point for research, but so far we have found no signs of life other than on Earth.

The search for extraterrestrial life is motivated and driven by the profound conviction that the question must be tackled now, and that we must find an answer soon. This belief is the result of the evolution of Western thought over the last few centuries. Earth, which before Copernicus seemed to be the center of the universe, lost its privileged position to the Sun, which in turn found itself in the unenviable situation of being a rather ordinary star in our galaxy, itself lost among billions of other galaxies in an unmeasurable universe. Is not the next step that of thinking that earthly life is not exceptional either?

While we wait for irrefutable astronomical evidence which would channel future research, all hypotheses about life in the universe are permitted. The special trademark of this research — the lack of evidence, the initial conceptual vacuum — makes it all the more interesting and gives it an added dimension. Anything goes, and research on life in the universe is going off in many directions at once. Some researchers take an inventory of the building blocks of life in the interstellar medium. Others attempt to estimate the probable number of sites favorable to life in our galaxy, then consider the consequences of the likely presence of life or even of technological civilizations elsewhere in the universe.

They do not hesitate to look to science fiction for inspiration when trying to explain the universe's silence. The job becomes multidisciplinary; for example, anthropological studies of Polynesian societies stimulate studies of the colonization of the Milky Way. Aren't the stars of our galaxy comparable to islands in the Pacific Ocean, separated by huge stretches of water whose crossing endangered the lives of colonists? Others, more pragmatically, are searching for planetary systems around the stars, or are listening to the universe, hoping to intercept a message or simply an artificial signal that would indicate intelligent activity elsewhere than on Earth.

The reader too can contribute to this research. A friend with no special scientific training suggested that viruses which cause epidemics on Earth could come from outer space, carried by dust — why not? This book is, first of all, the result of a thorough examination of bibliographical sources and many discussions with colleagues in astronomy and other sciences. But it was also enriched by questions and remarks put forth by the public at various conferences. It is indeed the public who showed me what was at stake, forced me to extend the problem beyond the limits of astronomy, and convinced me to abandon my scientific "objectivity" in order to take a stand and contribute to the debate. This sort of general examination is necessary for a preliminary survey of the subject and to give us the conceptual tools needed for the job. We have to generalize the familiar notions of "civilization," "communication," and even "life," the only forms of which we know are on Earth. We must try to put our existence into perspective, to examine ourselves from the outside.

Curiously enough, one of the greatest merits of this research is that it teaches us about ourselves. We must look at our planet and its inhabitants through a stranger's eyes. This examination from an unusual point of view can teach us much about ourselves, our limits and our particularities, but also what is universal about us. It is a way of knowing ourselves better by redefining who we are in a wider context.

Life on Earth

1

Comets and the Origin of Life

This first living cell, which carried the seed of all love, all thought and all music, and this human body, so beautiful when it is beautiful, and all that it has inspired, would be enough to make me go down on my knees.

—François Mauriac, *The New Notebook,* 1958–60

Lost in a foreign country, a traveler tried to find a city. The only thing he knew about it was its name. The city in which he lived was located in a fertile valley on the shores of a river, so he took the roads that followed rivers. Shouldn't we follow his example and, before looking for life in the universe, examine the only experience we know of — that of life on Earth — to find a starting point for our investigation? Are the physical and historical conditions to which we owe our existence common enough to be found elsewhere in space and/or time?

The inventory of vital resources starts here on Earth. We have to study the history of life on our own planet and to examine as precisely as we can anything that could help us determine its exceptional, or else its universal, character.

Three questions have to be asked: How did life start on Earth? Was the evolution of life as far as man the result of chance? Is our destiny one of progress, on Earth or in space?

Even without having thought much about it, we all have a preconceived idea about life in the universe. It is clear that the answers we will try to give to these three questions will either reinforce or disturb our initial convictions. If we are condemned to disappear in the short term, the chances that extraterrestrial civilizations have escaped the same destiny and are now communicating in interstellar space would seem to be slim. Furthermore, if we owe our existence to extraordinary historical circumstances, evolution in the direction of more and more complex organisms is probably not inevitable and we may be a rare, if not unique, species. Let me set aside these two possibilities for the moment and try to answer the first question.

How did life start on Earth? In fact, neither of the two currently accepted theories resolves the question. According to one theory, life is a fundamental component of the universe and has always existed. It developed on Earth from germs brought by dust from outer space. This is the *panspermia* hypothesis. According to the other theory — that of *spontaneous generation* — life emerged on Earth from inert matter, and it probably does not exist elsewhere. Between these two extreme points of view, numerous scenarios can be contemplated in which comets and meteorites might play a part.

The word *panspermia* was coined at the beginning of the twentieth century by the Swedish scientist Svante Arrhenius. But the idea that the seeds of animal and vegetable life are widespread throughout the universe and take root wherever and whenever the environment is favorable is much older. Around 500 B.C., the Greek philosopher Anaxagoras defended such a hypothesis. The idea remained latent until the nineteenth century, when it regained popularity after Louis Pasteur's famous fermentation experiment[1] delivered an apparently fatal blow to the hypothesis of spontaneous generation.

Two methods were suggested to explain the diffusion of life-carrying germs through the Galaxy.[2] According to the *lithopanspermia* hypothesis, upheld by Hermann Helmholtz and Lord Kelvin around 1870,the germs are carried by meteorites. According to Arrhenius' theory of 1908, the germs are pushed by the radiation pressure of stars (*radiopanspermia*).

1. Pasteur showed that fermentation was caused by microorganisms and could not happen spontaneously.

2. On the use of the term *Galaxy*, see the appendix.

Both means of transportation pose many difficulties. In the case of radiopanspermia, the diffusion speed of the germs is very slow — too slow, if we consider the distances that have to be covered to fertilize the whole Galaxy. Since the beginning of the universe, they would have covered only 6,000 light-years — one-eighth of the way.[3]

If the germs are transported by meteorites, difficulties arise at the moment when they enter the atmosphere. The heating up of the vehicle and its subsequent crash on the ground might kill the germs. This is true for meteorites weighing between a millionth of a gram and 10 grams. Organic material inside heavier meteorites can survive, since atmospheric friction heats up only a one-millimeter-thick outer layer of the vehicle. Very heavy meteorites (a ton or more) are completely destroyed on impact, because the atmosphere cannot slow them down sufficiently. At the other end of the mass range, fine interplanetary dust particles (up to 50 microns in diameter) land on Earth gently. The heat they absorb from friction in the atmosphere is usually not enough to make them melt. It is currently estimated that 8,700 tons of unmelted material land on Earth daily, 10 percent of it being intact organic carbon. The rate must have been much higher in the first million years of the Earth's history, before all the debris left over from the formation of the solar system was swept away. The net result after a hundred million years should be a thin layer of one thousandth of a gram of organic matter per square centimeter on the Earth's surface.

An American astronomer, David Brownlee of the University of Washington, gathered more than a thousand of these particles in the stratosphere at an altitude of about 20 kilometers. He used a plastic sheet covered with a layer of silicone oil and attached to a U2 plane, famous in other circumstances. He also dredged marine sediment 5,000 meters deep in the middle of the Pacific Ocean using a magnetic rake. Although these methods were efficient, they allowed Brownlee to collect only certain types of particles: ferrous or very small ones.

3. Our Galaxy (the Milky Way) has a radius of 45,000 light-years. A light-year is the distance traveled by light in one year at a speed of 300,000 kilometers per second.

A French group led by Michel Maurette of Orsay gathered a more complete sample of extraterrestrial material in Greenland in July and August of 1984 and 1987, and in Antarctica from January to March of 1988. In such environments, the particles also had the advantage of being well preserved, being far from the main sources of corrosion. In Greeenland, the micrometeorites were extracted from the black sediments found in the bottoms of shallow lakes near glaciers. In Antarctica, the team melted several hundred tons of ice to obtain their sample. Analysis of these samples showed that they had not undergone significant heating upon entering the atmosphere. About 30 percent of the micrometeorites had not melted, including some unexpectedly large ones (up to one millimeter). Dust therefore seems better adapted than meteorites for introducing the germs of life on Earth.

The voyage through the interstellar medium, an environment particularly hostile to terrestrial life, has risks of its own. The physical conditions are extremely varied. Very dense and cold regions where the temperature is close to absolute zero (–273°C) are separated by huge regions of very hot (several million degrees) diluted gas. The medium is streaked by lethal radiation: cosmic rays, x rays, ultraviolet radiation, and electron bombardment can all destroy life.

Laboratory experiments conducted by J. Mayo Greenberg in Leyden have allowed him to quantify the chances of survival of bacteria subject to ultraviolet radiation in the interstellar medium. For the bacterium *Bacillus subtilis*, the lifetime is of the order of 150 years. Only one bacterium out of 10,000 can survive beyond 1,000 years. However, clouds of interstellar dust can form an efficient shield, since they reduce the ultraviolet flux by at least a factor of 1,000. Furthermore, atoms and molecules deposited on bacteria form a protective crust which extends their life by tens of thousands of years, long enough to allow them to travel from one planetary system to another.

Cosmic rays present a much greater hazard for traveling germs than ultraviolet radiation, because they are much more energetic and because there is no efficient protection against them, except for a magnetic field like that which surrounds the Earth. The chances of encountering showers of cosmic rays are difficult to estimate, because their origin is not well known, but they certainly further reduce the lifetimes of germs.

The hypothesis of *directed panspermia* partly solves the difficulties. In this hypothesis, life would have been artificially implanted on Earth by intelligent beings from elsewhere in our Galaxy — beings a few million years ahead of us in evolution. This is the hypothesis advanced by Francis Crick. One can also imagine an accidental insemination by members of an advanced civilization who stopped on Earth during an interstellar journey and contaminated it by leaving scraps of their picnic.

But the real fault in the panspermia theory, beyond the problems of transportation and survival, appears when we draw all the consequences of this theory. Even if it explains how life could have emerged on our planet, it evades the problem of the origin of life itself. Either life necessarily precedes life, and consequently life has always existed, as has the universe, or life was born somewhere in the universe — but why not also on Earth? This theory establishes a discontinuity between living matter and inert matter. It implies that the properties of life cannot be understood by our scientific methods because these properties cannot be reduced to their chemical components.

Before excluding the possibility of the germs of life traveling through interstellar space and coming to stimulate life in privileged locations, we must examine the possible vehicles in greater detail to find out whether the germs left any traces on them. We must also enlarge the notion of germs to include, besides the recognized definition of a living organism, that of an essential element of life at a more or less complex level (organic molecule, amino acid, protein, etc.).

Among the meteorites gathered on Earth, only the carbonaceous chondrites are relevant to the origin of life, because they contain up to 4 percent carbon and because they contain organic compounds. The problem is to make sure that these compounds were not the results of terrestrial contamination at the time of impact or later.

Standard techniques, such as chromatography and mass spectroscopy, were used to analyze various meteorites: Murchison

(Australia) in September 1969, Murray (Kentucky) in 1951, and Mighei (Odessa, USSR) in 1966. These studies revealed the presence of biological amino acids, and others that are normally absent in proteins. Amino acids in D and L configurations were discovered in almost equal amounts, whereas living beings are formed almost exclusively from L-configuration amino acids. (The D and L configurations are two different arrangements of asymmetric molecules, one being a mirror image of the other.) All this leads us to believe that these amino acids were produced by non-selective — that is, non-biological — syntheses in interstellar space. (A meteorite which fell near Orgueil, France, in 1864 was examined by Marcelin Berthelot and Louis Pasteur, who tried, without success, to make possible extraterrestrial microorganisms grow by cultivating a sample of the meteorite's nucleus with organic matter.)

Antarctica is an ideal site for hunting meteorites; since they land on frozen soil, they are easy to find and are not contaminated by earthly organic matter. A total of 4,000 meteorites, 40 of them carbonaceous chondrites, have been brought back from Antarctica. Detailed analyses of two of them by a Japanese team confirmed the presence of amino acids of extraterrestrial origin.

Comets and interstellar dust are other potential vehicles for life-carrying germs. Like meteorites and interplanetary dust, comets contain high percentages of organic elements. Comets were probably formed in the protoplanetary disk at the same time as the planets. Interstellar dust constitutes an important part of the protostellar nebula where this disk was formed.

A comet does not have to collide with Earth for the latter to be contaminated by the comet's debris. Indeed, the dust ejected by comets as they pass near the Sun forms long trails through which the Earth passes periodically. This is the origin of our regularly recurring meteorite showers such as the Perseid and the Geminid. Judging by their trajectories, large meteorites (more than a kilogram) do not originate in the same place as dust and comets. They probably come from the asteroid belt — the thousands of small planets and large rocks trapped between the orbits of Mars and Jupiter.

9
Comets and the Origin of Life

(Observatoire de Haute-Provence)

Comet West. This photograph, taken at Observatoire de Haute-Provence (France) in 1976, shows the comet about ten days after its passage near the Sun, just before its core breaks into four pieces. The spiraling filaments visible in the plasma tail result from the interaction between the solar wind and gases ejected from the comet's core.

Comets are readily compared to dirty snowballs, made of icy and more or less complex molecules, of dust grains, and of crystals with gas trapped inside. Their exact composition is not yet known, but it is supposed that the molecules composing the core are at least as complex as those observed in the comet's trail. They could even be amino acids. However, the particular physical properties of comets, such as low mass, low temperature, and absence of atmosphere, of liquid water, and of lithosphere, make more complex biological syntheses unlikely.

European, Japanese, and Soviet space probes were sent to Halley's Comet in 1986. Their close encounters with the comet provided scientists with new data to test the hypothesis of panspermia. The European probe Giotto detected in the comet a polymer of formaldehyde called polyoxymethylene, which is formed of long chains of identical molecules. This is the most complex organic compound ever discovered in space. Fred Hoyle and his team in Cardiff have interpreted certain properties of the comet's infrared light in terms of bacterial grains, but this view is not widely accepted among specialists.[4] On the other hand, there is no evidence for amino acids in the data collected by the Soviet probe Vega as it flew by the comet.

It is still too soon to decide if comets, meteorites, and dust are at the origin of life on Earth. It is nevertheless certain that they all played important roles in the scenario of the appearance of life on Earth. These objects helped shape the surfaces of the inner planets, as the numerous craters on the Moon, on Mars, and on Mercury testify. They enriched Earth's atmosphere with life's building blocks, such as hydrogen, carbon, nitrogen, oxygen, and sulfur, and perhaps also with organic molecules. Models of planet formation predict that the influx of comets and interplanetary debris on Earth was much heavier in the past, and very intense up to 3.9 or 4 billion years ago. The appearance of life probably coincided with the end of this period of intense bombardment.

The idea that life appeared spontaneously from inert matter is as old as humanity itself. It is found, almost identical, in Genesis,

4. Hoyle's controversial hypothesis of bacteria in space is discussed further in chapters 5 and 12.

in the Rig Veda, in the philosophy of Anaximander (a Greek of the fifth century B.C.), and in many primitive myths. Newton and Descartes believed in the idea in the seventeenth century.

Considering the difficulties facing the panspermia hypothesis, in 1911 Paul Becquerel said that life was "a humble daughter of the Earth." During the 1920s, Alexander Oparin in the Soviet Union and John Haldane in Great Britain tried to reconstitute the history of the terrestrial atmosphere and the physicochemical conditions which preceded life. More recently, the paleontologist Edouard Boureau was able to retrace cellular evolution from bacteria to the first nucleated cells by studying Precambrian fossils. All these studies suggest that life arose on Earth from inert elements, over a period of time significant by human standards.

Many biologists and biochemists now believe that under special conditions, such as those in the Earth's primitive environment, life could be the final result of chemical reactions catalyzed by thunderstorms and the Sun's ultraviolet radiation. The chemical and biological research of Ilya Prigogine also goes in this direction: "Irreversible phenomena permit us to show that the roots of life plunge more deeply than we had thought into the inanimate."

On their side, astronomers who are searching for extraterrestrial civilizations are convinced that life must necessarily appear anywhere in the universe as soon as a sufficient level of complexity of chemical reactions is attained.

Pasteur's experiments would be valid only in a limited temporal context, and would not exclude the possibility that carbon compounds transform slowly and gradually into living systems over several hundred million years.

The Earth and the solar system formed about 4.5 billion years ago, 10.5 billion years after the birth of the universe. The original surface of the Earth was probably too hot for water to remain liquid. As the temperature of the Earth's crust progressively decreased, the oceans and the primitive atmosphere (mainly composed of water vapor, methane, carbon dioxide, and ammonia compounds) appeared, after volcanic and other phenomena allowed gases to escape from the planet's crust. It is in this reducing (i.e., oxygen-free) atmosphere, very similar to those of Jupiter and Saturn, that life would have started — or, as the panspermia hypothesis asserts, developed from a space-borne germ.

According to Boureau, life could even have started at several points in space and time during the Precambrian period, with varying amounts of success. "There were numerous physicochemical attempts," he writes, "but not all were successful." If this hypothesis is correct, it has important consequences. If life appeared only once on Earth in 5 billion years, the generating principle is rare indeed. However, if life appeared at several distinct periods, probably under different physical and climatic conditions, in each case it followed the same principle of evolution. If this second hypothesis is correct, there is a strong possibility that quite similar life may exist elsewhere in the universe.

The most ancient sedimentary rocks we know, located in Isua, Greenland, are about 3.8 billion years old. Analysis of these rocks seems to indicate[5] that life already existed at that time; however, we cannot be certain because of the intense metamorphism undergone by these rocks. What is certain is that biological evolution was extremely slow during the 3.4 billion years of the Precambrian period. The fossils dating from that period are extremely simple organisms: anaerobic bacteria (the earliest from 3.2 billion years ago), and unicellular blue algae (which still exist). Spheroids — larger organisms from which the first eukaryotic cell (cell with a nucleus) is thought to have evolved — date from the end of the Precambrian era, about a billion years ago.[6]

Evolution toward the first eukaryotic cells is well explained by the fundamental notion of symbiosis. Probably for defense against an aggressive environment, primitive earthly organisms first regrouped by symbiosis and later surrounded themselves with a membrane. Spheroids, for example, formed into very diverse bacterial colonies, surrounded by a common wall. The insertion of a foreign organism into such a group can appear at first like an

---

5. The ratio of light carbon (isotope ^{12}C) to heavy carbon (isotope ^{13}C) is abnormal; it was probably modified by biological synthesis, which shows a marked preference for light carbon.

6. The exact date has not been established because of the uncertain role (nucleus or germinative pore) played by the "black spot" of some spheroids during the Precambrian period.

infection. Then, after mutual adaptation, the new element participates in the life of the system and contributes to its evolution toward the eukaryotic cell. What does seem obvious is that fossils of that period are collections of spheroids which become more and more complex with time.

The first traces of a reproductive mechanism, an essential attribute of life, were discovered in some fossil cells, notably spheroids. These are the budding and the emission of substances from the cell, which can be interpreted as the first stages of mitosis, the most frequent form of cell division.

The earliest known multicellular organisms are the large populations of ribbons with regular morphology discovered in northwestern Canada and China. Their age is estimated to be 800 or 900 million years.

Even though life appeared early in the history of our planet, this first stage of its evolution was extremely long — over 3 billion years. The living world that we know today, with its multitude of species, really began only with the appearance of an oxygen-rich atmosphere, 500 million years ago. At that time, the atmosphere became oxidizing, one molecule out of five being oxygen. A layer of ozone formed in the upper atmosphere and considerably reduced ultraviolet radiation on the Earth's surface. This ended the stage dominated by simple forms of life.

This turning point in the evolution of Earth's atmosphere and of life deserves to be analyzed in detail. The living organisms themselves enriched the atmosphere with oxygen, through cyanobacterial photosynthesis. They modified the environment, maybe at their own expense, and opened the way for the development of a more efficient metabolism and thus for complex forms of life. We can thus imagine life developing in a particularly unfavorable environment, but perhaps progressively improving this environment and then diversifying itself.

However, it was only with the stabilization of the continental blocks and the reduction of volcanic activity that the net result of photosynthesis became positive and the atmosphere became oxidizing. Therefore, the appearance of more elaborate forms of life depended principally on tectonic activity, and it could have occurred much earlier — or not at all, if the Earth's crust had developed at a different rate.

This intrusion of the imponderable into life's evolution merits thought. If greater or more precocious volcanic activity had shortened the time separating the first bacteria from *Homo sapiens* by several hundred million years, we would now be at an infinitely more advanced stage of social and technological evolution. Perhaps we would already have colonized planets around neighboring stars and achieved some of the dreams of our best science fiction writers. This opens up the prospect of extraterrestrial civilizations that are technologically much more advanced than we. This is the first clue that the analysis of life's emergence on Earth brings us about the possible existence of universal life. There does not seem to be any incompatibility between the length of life's evolution on Earth and the existence of more advanced extraterrestrial civilizations in planetary systems comparable to ours. But such civilizations are only a possibility, and we could destroy ourselves before reaching that advanced stage.

Even though the first steps of the evolution of life and its attendant risks are now practically known, the origin of life itself remains a mystery. In what "warm primitive soup," in which particularly favorable environment, could the first spark of life been born? Prebiotic chemistry in Earth's primitive atmosphere is essentially composed of hypotheses.

According to Oparin and Haldane, it would have been quite similar to the present atmospheres of Jupiter, Saturn, and Titan: mostly methane, ammonia, nitrogen, and water. In their opinion, one had to look to oceans for the "warm, primitive soup," that alchemist's broth in which life could have been born. But the idea of a reducing atmosphere is not completely accepted. Some think that there could already have been appreciable quantities of oxygen. Others believe that the primitive atmosphere was similar to that on Venus and Mars, containing only carbon dioxide, nitrogen, and water — a medium rather unfavorable to the production of organic molecules. Those would have come from space.

Recently, Californian and Scottish biologists advanced the hypothesis that clay could have been the famous "primitive soup."

They believe that clay is a more favorable environment for prebiotic chemical reactions. Clay stocks energy emitted by natural processes such as radioactivity in the form of electrons. Molecules contained in the oceans could come into contact with the clay matter on the shores during high tide. Then the clay would catalyze the chemical reactions by freeing these electrons under the effect of tensions caused by drying. The clay thus acquires the nature of a proto-organism which would act as a model for future living beings. In fact, all it lacks is reproductive ability. Perhaps Genesis should be taken literally after all!

Scientists acted as apprentice sorcerers by re-creating in laboratories what they believed to be the primitive atmosphere. They tried to create prebiotic molecules, amino acids, and the more ambitious ones even tried to create life. The first important experiment, by Stanley Miller in 1953, involved sending electrical discharges of 60,000 volts into a gaseous mixture of ammonia, methane, hydrogen, and water. A week of this violent treatment produced amino acids. This was followed by more experiments almost everywhere in the world. In France, Gérard Toupance and François Raulin used quite similar gaseous or liquid mixtures, and their sources of energy were electrical discharges, electrons, and ultraviolet radiation — all abundant energy sources in the primitive Earth — or else protons, gamma rays and x rays, or just heat. Of the twenty amino acids found in proteins, fourteen were obtained by synthesis in the laboratory. The components closest to those of life were produced experimentally by Sydney Fox in 1980. By heating amino acids to 170°C on lava, he obtained chains of several hundred molecules which, when dissolved in water, produced "microspheres" resembling Precambrian spheroids. In 1989, a team at the Space Research Institute in Moscow heated meteoritic material with energetic laser pulses to simulate the vaporization of meteorites as they hit the Earth. The gas mixtures that formed did indeed contain organic compounds such as methane, cyanhydric acid, and acetaldehyde.

But were the particular conditions used in these experiments realistic? The ultraviolet radiation used to synthesize amino acids could dissociate them before they have time to form more complex biological molecules. Could a catastrophic event like a comet crashing onto Earth furnish the necessary spark? Other energy sources for these reactions could be emanations from radioactive elements, shock waves from lightning, or meteor showers. Table 1 summarizes the possibles sources of energy in the primitive atmosphere.

Table 1
Energy sources in the primitive atmosphere.

Source	Energy (calories/cm^2/year)
Solar radiation (all wavelengths)	260,000
Solar radiation (ultraviolet)	570
Electric discharges	4
Cosmic rays	0.0015
Radioactivity (to depth of 1 km)	2.80
Heat from volcanoes	0.13
Meteoritic impacts	0.10

Another direction has been taken by the scientists at Besançon Observatory in France who study photochemical models of the primitive atmosphere by computer simulation. The evolution of the atmosphere and the production of prebiotic molecules are studied by means of a model incorporating everything known about the primitive atmosphere: chemical composition, temperature and pressure, and the types of reactions that any components present might produce. Such a theoretical approach is safer than an empirical laboratory method where the chemical constituents resulting from reactions are difficult to isolate and identify. Some scientists suspect that the compound molecules discovered in the laboratory were actually produced by the process of identification, after the reaction itself. Computer models do not present this difficulty and allow us to better understand how those molecules form by following the reactions step by step. Another advantage of this method is the possibility of studying the global evolution of the atmosphere, taking all the known reactions into account, whereas laboratory experiments treat reactions separately.

Rather than follow evolution toward the first living cell, genetic biologists have gone back in time toward that cell. The results are surprising. The analysis of molecules carrying life's genetic code allowed them to build the genealogical tree of living species. First they had to identify the sequences of amino acids which constitute these molecules; they then identified the series of sequences common to several species and those that, on the contrary, are incompatible with a common origin, thus finding all the affiliations. In this way, James Lake, a biologist at the University of California, went back to what he thinks is the first living cell by comparing the sequences of ribosomes (one of the components of living cells) in the five most primitive species of bacteria now extant. The last ancestor common to all living organisms would resemble *eocytes*, bacteria which live in sulfurous geothermal springs at temperatures close to 100°C. Such an environment resembles neither the original warm soup nor even the clay already suggested as a possible environment for the first living cells.

The discovery of numerous oases of life around hydrothermal sources below 2,000 meters in the Pacific and Atlantic oceans in the past 15 years lends credit to the possibility that life indeed originated in a very hot environment. Such hydrothermal sites provide a unique combination of gases, heat, and water, which were present in nearly the same proportions in the primitive oceans. There are, however, some difficulties with the view that life originated deep in the oceans. For example, the basic amino acids become very short-lived at temperatures above 200°C, even under high pressure. The future exploration of life in the deep-sea environment by special submarines such as Alvin may shed some light on the question of where life originated.

Another area remains to be studied: the selective factors which on Earth direct the transformation of molecules into amino acids. There are only twenty amino acids in living organisms, almost all in L configuration.[7] Remember that two molecules of the same composition but in L and D configurations are like gloves for the

7. If biological amino acids are of L configuration, other molecules such as membranes of bacteria are in D configuration. In fact, most living molecules are optically active (i.e., either L or D). The reason for and the origin of this activity are still unclear.

left and right hands. Miller's experiments produced all kinds of amino acids, in both L and D configurations. The mixtures of amino acids produced by these experiments are more like carbonaceous chondrites than a biological medium.

Synchrotron radiation emitted by supernovas could furnish an explanation for this selective phenomenon. Such radiation, caused by electrons spiraling at very high speed along the lines of a magnetic field, is strongly polarized (i.e., the radiated light waves vibrate in one plane). It is thus capable of inducing dissymmetrical, photochemical reactions and favoring the production of certain amino acids rather than others.

The Sun is another source of polarized light that could influence the production of asymmetrical amino acids. Sunspots are polarized as a result of the intense magnetic fields at the Sun's surface. Diffused daylight is also polarized through the dispersion of aerosols in the atmosphere. This polarization can be significant on sloped land, for example near mountain ranges. The terrain thus hides certain regions of the sky, and the average polarization in one day is not negligible.

Far from giving any conclusive answers, research in prebiotic chemistry only poses more questions about life and its origin. Life is not ready to give up its mystery. Analyses of meteorites and dust which have landed on Earth have turned up no obvious traces of extraterrestrial life. Knowing that the interstellar medium is hostile to life, I tend to prefer spontaneous generation to the panspermia hypothesis. But were the 3 billion years which went by before the appearance of the eukaryotic cell enough time for the appropriate elements to associate, after innumerable reactions and sterile combinations, in conditions which might give birth to life, and then to give it the cellular structure which is the basis of all living organisms?

Despite current uncertainties about life's origin, we must not rush into explanations that depend on extraordinary intervention, whatever its temporal or spiritual nature; this would only reinforce the exceptional character of life. We will find the answer at the telescope, while trying to discover planets like ours around stars like the Sun, or else while sending probes to look for life in the solar system (especially on Titan and Europa). The paradox of this first chapter is that the search for life in the universe can bring us more information about the origin of life on Earth than proceeding in the opposite direction would. Only the slowness of life's initial evolution on our planet allows us to hope for the existence of more advanced intelligent civilizations in the Galaxy. This indication is important because it has encouraged the search for artificial radio signals.

2

The Temporal Dimension of Life

Time is the deepest and most tragic preoccupation of human beings.
—Simone Weil, *Gravity and Grace*

Even though the origin of life on Earth remains a mystery, its history is known fairly precisely thanks to biology and paleontology. The driving force of this history — the central element without which we could not even call it history — is a remarkable phenomenon that astronomers find all over the universe. I am talking about evolution.

The idea of evolution was already mentioned in the preceding chapter. It is indispensable to theories and empirical research on the origins of life. The concept of evolution also explains the variety of species currently alive.

An inventory of life forms from Primary-era bacteria to man immediately shows that evolution went in the direction of ever more complex organisms. There were certainly some cases of regressive evolution. Cave-dwelling animals, for example, progressively lost the use of sight, which was of no use in a lightless environment. But in all cases, evolution is inevitable. An extinct species has never reappeared. This idea of irreversible evolution, which has not yet stopped disturbing modern scientific thought, is the true temporal dimension of life. It shows us that some events necessarily precede others, and thus, the direction in which time flows. I will even go further than Henri Bergson in his *Creative Evolution* and affirm that evolution *causes* this passage of time.

If we put our history into the more general framework of the universe, we realize that the evolution of life is just another aspect of the evolution of the universe. One of the most important innovations of modern astronomy has been the demonstration that the universe is not inalterable but, on the contrary, is continually expanding and evolving. From this point of view, life no longer appears as an exceptional dynamic process in a set universe, but rather as one of the aspects of the evolution of a universe that is constantly changing and becoming richer and enlarging itself.

In the 1920s, the American astronomer Edwin Hubble discovered that all galaxies, except those closest to us, are moving away from us — the greater their distance from us, the greater their speed. This was the starting point for modern cosmology, which is based on the hypothesis of an expanding universe. The fact that all those galaxies are moving away from us could mean that we are at the center of the universe. However, the true explanation seems to be that all galaxies are moving away from one another. To understand this idea, we can compare the universe to rising dough, the galaxies being raisins in the dough. As the dough swells, the raisins move further away from one another. This phenomenon of expansion has since been confirmed by precise measurements of the radial velocities of tens of thousands of galaxies, although the rate is much lower than Hubble estimated.[1]

The discovery of cosmic radiation at 3°K (i.e., – 270°C) confirmed the idea of an expanding universe.[2] Cosmology predicts that the universe was born in a primordial explosion, the "Big Bang," at a temperature of 100 billion degrees. The remainder of this event is cosmic radiation. This radiation, detected mainly in the radio and infrared domains, is very weak and perfectly uniform throughout the universe.

Like the birth of life, the birth of the universe remains a mystery. We cannot imagine the primordial explosion, nor can we understand it through physics, because it is not an event. This

---

1. The exact value of this rate, Hubble's constant, is the object of much controversy among astronomers. It is probably about 25 kilometers per second per million light-years.

2. This radiation was detected in 1965 by two researchers at Bell Telephone Laboratories, Arno Penzias and Robert Wilson.

non-event does not belong to space-time; it is one of its frontiers. It is beyond the domain of scientific investigation. And yet the origin of the universe and its causes constitute one of the fundamental questions of science. But science only begins in the billionth of a second following the Big Bang.

Quantum physics could offer us a new angle from which to look at the birth of the universe. The creation of a particle in an intense electromagnetic or gravitational field is spontaneous and unpredictable. The field is necessary but is not the cause. If the laws of quantum physics allow particles to create themselves spontaneously, why could they not allow space-time to appear in the same way, for no reason, but simply by fluctuation from another state? For this to be possible, however, these quantum laws would have to exist even in the absence of the universe. This creates a basic philosophical difficulty, because of the notion of an observer in quantum physics: How can an observer exist outside a universe which, by definition, encompasses everything?

The expansion of the universe is the first cause of its evolution. After expanding for about 100,000 years, the universe had cooled enough for the components of matter to combine and form atoms of hydrogen (80 percent) and helium (20 percent). At this same time, the cosmic radiation whose traces we can still observe was emitted. Because of gravitational instability, the matter which until then filled the universe uniformly fragmented and contracted under the effect of its own mass. A hierarchical structure appeared in the universe, with stars its elementary building blocks. These stars concentrated into open (several thousand stars) or globular (up to several million stars) clusters, constituting galaxies (hundreds of billions of stars), which in turn group to form clusters of galaxies (several thousand members), which form superclusters, which form clusters of superclusters, and so on.

At the bottom of the hierarchy, a star's life is a constant struggle against gravitation, which suffocates it. The gas cloud from which a star is born heats as it contracts. When the temperature in the center of the cloud reaches 4 million degrees, nuclear burning of the hydrogen begins and produces intense luminous energy, which balances the force of gravitation and spreads out into space. A star is born. When all the fuel is used up, contraction starts again; it continues relentlessly until the temperature reaches about 100 million degrees. Now helium nuclear burning takes over and feeds the stellar furnace, producing carbon, oxygen, and nitrogen. These elements diffuse into space by supernova explosions (certain stars explode at the end of their lives) and also by more progressive means, such as stellar winds. The interstellar medium is thus enriched and new generations of stars, as well as planets and life, are formed. Some 15 billion years after the Big Bang, the cosmic proportions are 76.5 percent hydrogen, 21.5 percent helium, 0.82 percent oxygen, 0.34 percent carbon, and even lower percentages of the heavier elements.

This simplified scenario of a star's life reveals a second mechanism by which the universe evolves, one much more powerful than expansion. It is the progressive complexification of matter. Stars produce atoms of greater and greater complexity. At the end of a star's life (or even during its lifetime, if it is luminous enough to emit a stellar wind), these atoms disperse in the interstellar medium and assemble as molecules, themselves more and more complex. In this sense, we are born from stars. We are literally stardust, which life will complicate irreversibly, thus continuing evolution, which puts us into the current of time. This evolution is the very essence of time; it shows us the direction of the arrow of time.

Let us examine this idea more closely. Our lives take place in time, inexorably. We cannot go back in time. We realize that time goes by because we remember the past, but we do not know the future. This is a subjective perception, but it reflects a universal reality: everything in the universe takes place in time.

Scientists have identified three categories of physical phenomena in this temporal direction. The first, the *historical arrow*, is defined by the accumulation of information in our memory and in matter. Many geological phenomena have left traces on the surface of the Earth. The process of biological evolution from the original cell to complex organisms and the enriching of the interstellar medium by heavy elements both show the direction of time.

The *thermodynamic arrow* is linked to irreversible phenomena in physics and chemistry. Beside the orderly and ever more complex organization of life and of stellar systems, there is, paradoxically, increasing disorder through the generation of entropy.[3]

This definition presents a conceptual difficulty: the microscopic phenomena which increase entropy are perfectly reversible, whereas the dynamical laws which govern them remain valid if the course of time is reversed. We thus have to imagine that at the beginning the universe was in a very special and very orderly state. Furthermore, this definition is not perfect. When the drop of wine has dispersed completely in a glass of water, the mixture is perfect. We cannot establish any further evolution. The thermodynamical arrow won't show anything more.

The third arrow, *cosmological time*, is linked to the expansion of the universe. If the universe is infinite, expansion will continue indefinitely. If, however, the universe is closed, expansion will stop one day and contraction will take its place. In this case, at two distinct moments the universe will have the same radius, but the acceleration will be different. Our universe must be just at the limit of these two geometries, but we don't know on which side! This cosmological time is probably the closest to absolute time as it was imagined before Einstein — that is, time measured by a universal clock, the same for everyone.

The evolution of matter and life, the growth of disorder, and the expansion of the universe are three categories of phenomena which, on different scales, generate the flow of time. The latter is a consequence of the progressive construction of the universe, and of the disorder and wear and tear which accompany this construction. By their evolution, the universe and living beings achieve and continue a project. But here science is silent. It answers the *how?* but not the *why?* Would time exist in the absence of irreversible evolution of matter in the universe? Probably not, but we cannot go more deeply into this endless theme here. The object of what may appear as a digression about time is simply to show the similarity between the history of life and that of the universe.

3. Entropy is a measure of the degree of disorder in a system. In an isolated system it can only grow with time.

It is an irreversible evolution that engenders the fourth dimension of space-time.

Up to this point I have considered the evolutionary character of life as an acquired fact. But even though evolution is no longer in doubt, this was far from evident in the past. As a working hypothesis, it had to wait until the beginning of the nineteenth century and the writings of Lamarck. Darwin specified the new theory only in 1859, after a voyage to the Galapagos Islands, whose very particular fauna suggested to him an explanation for the diversity of the living world.

The variety of life had for a long time been considered the result of spontaneous creation with no natural cause. This is still, in fact, the creationists' point of view. We have to look for the driving force of diversification through a microscope. Life on Earth diversifies through genetic mutations — random modifications which affect the hereditary constitution of certain cells. An entire organism can be modified by these mutations.

Most mutations do not allow the individuals involved to adapt better to their environments, because in a given context the possible number of mutations is always limited. There are now more than a million species living in the world, but about 500 million have disappeared because they could not adapt quickly enough (or well enough) to changes in their environments. Among humans,it appears that half of all conceptions abort during the first 3 weeks of gestation and are not even noticed. These spontaneous abortions are due to defects in the chromosomes of the embryo. They are probably the best response of the present human organism to stresses most likely caused by pollution of the environment.

Genetic mutations explain how the living world diversifies, but they are not the first cause of its evolution. This cause must be looked for in population genetics. Ever since Darwin, the principal mechanism used to explain the evolution of living organisms is natural selection. According to François Jacob, it is the result of two constraints that are imposed on all living organisms: the need to reproduce and the need to interact permanently with one's environment. The same parents produce completely different individuals, and the populations of different species vary throughout time. Environment takes care of the rest. Various constraints, such as geographic isolation, contribute to the divergence of species. Natural selection directs evolution, which is a slow process, from what already exists. "Novelties," writes Jacob, "come from previously unseen association of old material. To create is to recombine."

Natural selection does not follow a well-established plan that would necessarily lead to intelligent organisms. Any attempt to reduce evolution to simple and precise laws can only fail, because the rules of evolution are specific to each level — chemical, biological, or social. Natural selection seems to work in jolts, according to circumstances and conditions. The historical circumstances which condition the interaction of living organisms thus play an important role in their evolution, as do the constraints of the physical environment in which they live. Astronomical events can have dramatic consequences, sometimes discernible only after several million years, as they disturb the course of evolution and completely change the relations among different living species.

The most obvious constraints are climatic conditions, as when reproduction requires a wet environment or an atmosphere containing oxygen. Other social or economic constraints appear as living organisms become more complex. They then become more sensitive to historical circumstances: a species, animal or vegetable, can develop at the expense of others, and force them into regions where the conditions for maintaining life are more difficult. There they adapt or disappear.

Identical constraints can thus produce similar results along differing paths. For example, eyes are common to numerous living species because they answer the same need — that of seeing where one is going. However, the details of their mechanisms can vary considerably from one species to another. Forty independent cases of eye evolution are known. Wings are found on insects, fish, mammals, and prehistoric reptiles. Certain organs are found more often under certain conditions: webbed feet on aquatic animals, smooth fur on burrowing animals, a prehensile tail on tree-dwelling creatures. This phenomenon of the convergence of very different species was established as a principle by biologists. At the end of this chapter, I will extrapolate it to life in interstellar space.

Despite the diversity of living species, all possibilities have not yet been exhausted. All the same, it is remarkable that on all the continents, including Australia, animal species all developed the same organs, the same morphologies: legs, wings, a head with two eyes, and so forth. Why, for example, aren't there any animals with wheels instead of legs? The wings of certain butterflies have patterns on them to discourage birds from eating them; why not, rather, an unappetizing taste or smell? Embryology might be able to supply answers to these questions. The laws of physics and chemistry also limit the possibilities.

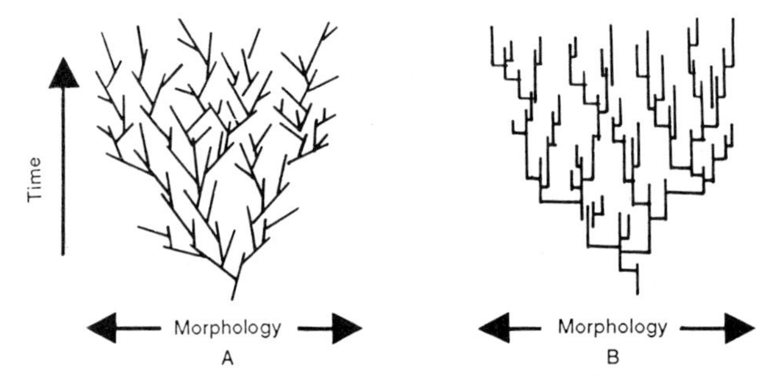

Two diagrams of evolution. In the hypothesis of gradual evolution (left), the various living species represented as branches of the evolutionary tree separate progressively from the trunk. In the hypothesis of punctual evolution, the separation is abrupt. *(Adapted from H. Tintant)*

Natural selection is far from being the unanimous choice as the driving force of evolution. One of the basic concepts of Darwinism is "gradualism" — the idea that evolution is a slow accumulation of genetic changes. In the past decade, it has been especially contested by the American paleontologists Stephen Jay Gould and Niles Eldredge. They claim that evolution would more likely happen abruptly, by "punctuated changes," after long periods of stability ensured by a high population of the concerned species. The changes would concern a very small marginal population which, under specific circumstances, develops disproportionately until it competes with the original population. In this "punctualist" theory, evolution is the result of sudden innovation rather than of slow adaptation.

Punctualism has revised the idea of species. Darwin thought admitting that species existed as individual entities was a return to fixism, in which species were collections of living organisms which were quite similar at a given moment and which reproduced among themselves. In fact, species show a surprising resistance to change. Some trilobite fossils, for example, show no changes for millions of years; then suddenly, over 5,000 to 50,000 years, there appear striking changes, such as the reduction of the number of lenses in the eye. Genetic changes accumulate without noticeable effect until they reach a certain threshold, after which the features of the species change dramatically.

The evolution of cephalopods (small mollusks, including ammonites) seems to combine gradualism and punctualism. An initial phase of innovative evolution, during which distinctive features of the species are elaborated, is followed by long periods of adaptative evolution in which the species gradually responds to changes in the environment.

Another anti-Darwinian theory of evolution has been suggested by Kinji Imanishi. In Darwin's theory the individual plays a very important role, and natural selection implies competition among individuals in producing more offspring. For Imanishi, there is coexistence, not competition. The survival of a species would be the result of chance rather than selection. Evolution is not a question of a few individuals, but of entire species that change when the time for change has come. Thus, Imanishi's theory is similar to the theory of group selection.

The central point of Imanishi's theory is segregation of the habitats of species, accomplished in a peaceful manner. The principles of cooperation and harmony in the living world and the priority of the group over the individual reflect the philosophy of Japanese society, while ideas of individualism, competition, and selection are those of Western society. Imanishi's theory of evolution is a poetic one. Unfortunately, it has no place in a scientific view of the real world, which is a place of struggle, competition, and disagreement.

Nevertheless, the notion of evolution is the common denominator of all these theories. The fixist idea of species is in total contradiction with the facts I have just mentioned. Creationists, who support fixism, have recently been much in the news for wanting their theory to be taught in American schools as well as the theory of evolution. They claim that the appearance of life on Earth is possible only with a Creator. Their arguments, based on religious dogma and biblical revelations taken literally, often play on the dissent among evolutionists. Of course, science only partially accounts for reality, and apparently divergent scientific theories coexist. But in our present state of knowledge, evolution remains an indisputable hypothesis. Even in the realm of religion, the theory of evolution is not necessarily invalid. Did not Pierre Teilhard de Chardin, the Jesuit philosopher and scientist, see God as the converging point of evolution?

If physical conditions are so important in the maintenance and evolution of life, what would have happened if they had been slightly different? Even though the answers might be reminiscent of medieval teratology, this is not a completely idle question. Perhaps we may find the answer to "What do extraterrestrials look like?"

If the Earth were smaller, its gravity would be weaker. On the Moon, things weigh only one-fifth as much as on Earth. Trees could grow to 150 or 200 meters. In order to eat the fruit on them, most animals would need necks like those of giraffes, or hind legs like those of kangaroos. But Earth could not be much lighter. If its mass were less than 70 percent of what it is, Earth would no longer retain its atmosphere. However, if Earth were more massive, gravity would force animals to the ground and they would have to crawl. To pull their heavy bodies, they would need a large skeleton, very solid limbs, a thick neck, and a powerful heart. But such superficial differences should not breed life fundamentally different from what we know. We can imagine that strong gravity would reduce the activities of life forms, especially their creativity. The development of intelligence and the evolution of culture would slow down. But perhaps the added difficulty of movement would stimulate adaptability in creatures and reflection in intelligent beings.

Let us be more daring and imagine for a moment that Earth has a different orbit in the Galaxy, independent of the Sun. Our daylight comes from a permanent glow in the atmosphere, and the heat necessary for life comes from an internal source. There would be no day/night alternation, and no seasonal or biological rhythms. More or less distant meetings with other celestial bodies would be rare on a human time scale and would provoke particular physical phenomena, without any periodicity.

Under such conditions, most of the regular and periodic processes that pattern our lives would not exist. Life could probably get used to this loss and create its own rhythm. However, the consequences for the development of physics would be dramatic. Indeed, science was certainly born from the observation of events that repeated themselves periodically, like the phases of the Moon or the rhythm of the seasons — their regularity suggesting the existence of laws or underlying explanatory mechanisms. In a universe with no apparent order, would we have dreamed of looking for these laws? Would we have developed physics, which may perhaps one day lead us to the discovery that we are not alone in the universe?

There is, therefore, a fundamental importance to the periodic motion of the Earth around the Sun, not only for the evolution of life but also for the evolution of thought. It is one of the moorings to which scientists can connect observations and experiments which help to determine general laws.

The universal nature of evolution makes the case for extraterrestrial life strong, if not convincing. But it is now clear that if life exists elsewhere, its form will certainly not be the same as on our planet. Even if life all over the universe was formed from the same materials as on Earth, even if it evolved on planets exactly like ours, the historical circumstances could not have been the same. However, my colleagues in the Soviet Union think quite differently. The parallel between the evolution of the universe and life has led them to extend one of the basic laws of physics — the principle of equivalence (or uniformity)[4] — to living matter. If

4. Physical laws are the same all over the universe. This principle allows us to account for most of the events observed in the universe.

there is equivalence in the evolutionary laws of living and thinking matter in the cosmos, the existence of intelligent extraterrestrial societies becomes inevitable.

Felix Tsitsin of the Sternberg Institute, in Moscow, even suggests the idea of a convergence of species in the cosmos. In interstellar space, where physical conditions are very similar, we could indeed expect all forms of life to develop in the direction of a convergence of morphological features. We could then imagine a community of extraterrestrial societies of various origins but made uniform by a common process, much as contemporary human societies are molded by Western civilization. This convergence would create affinities, and therefore reasons to communicate, among societies that had evolved along very different paths. Here is another reason for us to try to "join the club."

3

Evolution and Catastrophes

There is only chance because human interest is involved and because things happened as though man had been taken into consideration.
—Henri Bergson, *The Two Sources of Morals and Religion*

The appearance of man is a landmark in the slow evolution of life. Was this event inevitable, or do we owe our existence to a series of particular circumstances? If the latter hypothesis is correct, we are probably the only intelligent civilization in the universe. Such is the opinion of Jacques Monod. In *Chance and Necessity*, he writes that "man finally knows that he is alone in the indifferent immensity of the universe out of which he emerged by chance." It now seems clear that the forms and the paths that life takes in its evolution depend mainly on contingencies beyond its control. But instead of agreeing immediately with Monod's opinion, I will examine the catastrophes which marked this evolution and see if they played a determining role — if they really influenced the course of life.

Toward the end of the Secondary era, the animal world was dominated by giant reptiles: diplodocus, brontosaurus, tyrannosaurus, iguanodon, archeopteryx, and so on. These animals, known under the generic name *dinosaurs*, disappeared quite abruptly 65 million years ago at the time which now marks the limit between the Cretaceous period and the Tertiary era. With them disappeared 75 to 85 percent of plankton, and the ammonites which lived off it, as well as various marine and flying reptiles. (We must remember, for what comes next, that many kinds of terrestrial plants, freshwater vertebrates, snakes, mammals, and marine invertebrates were hardly affected.)

The disappearance of these huge animals left a hole in the animal world. This ecological gap has slowly been occupied by mammals. One of these, *Homo sapiens*, has come to dominate life on Earth. But what would today's terrestrial fauna be like if dinosaurs had not suddenly disappeared? Would man have been able to emerge, or would dinosaurs still be masters of all the other animals, as they were for 135 million years? To answer this basic question, we need to determine the events that could have caused this sudden extinction. It is indeed very possible that they could happen again, causing other extinctions, even that of man. The role played by chance in the evolution of life must be evaluated in order to establish whether the emergence of intelligence was inevitable or accidental.

At first, the extinction of the dinosaurs was thought to have biological causes. The disproportionate sizes of their tiny brains and their huge bodies would have made them vulnerable to other predators. Perhaps their nervous systems made them like automatons in comparison with mammals, which developed a minimum amount of consciousness, enabling them to adapt more flexibly to adverse conditions. Small mammals may have eaten their eggs more rapidly than dinosaurs could lay them. The fragility of the eggshell has also been suggested as a possible cause of extinction. We must also realize that dinosaurs, like all organisms in small number, were relatively vulnerable to any change in their environment.

The discovery of anomalies in the composition of the geological layers from that period has directed researchers toward other explanations. In 1980, Luis and Walter Alvarez and their collaborators at the University of California at Berkeley measured the amounts of heavy metals of the platinum group in their samples. They found quantities significantly higher than the Earth's average, especially in the cases of iridium and osmium. As the proportion was closer to those found in the interstellar medium, they attributed the difference to a meteorite or an asteroid about 10 kilometers in diameter which must have crashed on Earth. The excess of heavy metals was confirmed by other measurements taken at about sixty sites in the United States, Italy, Denmark, France, New Zealand, and China. The whole Earth seems to have been covered with a fine layer of dust several millimeters thick at the Cretaceous/Tertiary boundary.

The sediments also bear evidence that other catastrophic events, such as giant tsunamis[1] and a global forest fire, exacerbated the environmental stresses associated with the initial catastrophe. The presence of soot in large quantities has been reported in sediments from that period taken from five sites in Europe and New Zealand, and evidence of a giant tsunami has been found in Texas.

Other explanations are also possible. A supernova, a star exploding at the end of its existence, could have been the origin of this abnormal geological deposit. But it would have to have been very close to the solar system, and this is unlikely in view of the low density of stars in the solar neighborhood. Furthermore, a supernova would have produced plutonium, an excess of which was not found.

A period of exceptionally intense volcanic activity could have produced a layer of dust rich in heavy metals, although perhaps not one evenly distributed over the whole Earth. The basalt floors of the Deccan plateau in India date from this period. This volcanic activity lasted about 500,000 years, with more intense periods of 10,000 years during which 1.7 billion tons of sulfuric acid and 27 million tons of hydrochloric acid would have been emitted into the atmosphere.[2] The consequences of intense and prolonged volcanic activity would be catastrophic for life on Earth: sulfuric acid rains, reduction of the oceans' alkalinity, a 3° reduction in global temperature, and a 7 percent reduction in the atmosphere's ozone layer caused by the hydrochloric acid.

If associated with a general lowering of sea levels, these effects could explain a massive and progressive extinction. The survival of certain species may have been due to a high population, a

1. A tsunami is a tidal wave that occurs during earthquakes.

2. For comparison, industrial and domestic combustion emitted 120 million tons of sulfuric acid into the atmosphere in 1976.

nocturnal or subterranean life which would protect them from the Sun's ultraviolet rays, or a greater tolerance for temperature variations.

Finally, the Sun could have met a cloud of interstellar dust as it traveled around the Galaxy. It crosses a spiral arm of the Galaxy, rich in interstellar matter, about every 250 million years. Such a cloud between the Sun and the Earth would have absorbed sunlight and re-emitted it mostly as infrared light, thus dramatically modifying Earth's photosynthesis and affecting any life form that needed sunlight. But such a phenomenon probably would have lasted about 250,000 years, whereas the iridium layer corresponds to a maximum duration of about 10,000 years.

The hypothesis of a meteorite's impact seems to have been confirmed in the past few years by the discovery of microcrystals with metamorphic signs produced by shocks and very high pressures, at least 100,000 times the pressure on Earth. Such microcrystals were found in Spain, Denmark, New Mexico, New Zealand, and Italy. No volcanic eruption could have created such extreme physical conditions. Nor could it eject such large crystals (0.3 millimeters in diameter) high enough into the atmosphere for them to be found in sites so far apart.

The meteorite may even have been larger than first estimated. Oceanologists at the University of Rhode Island measured the amount of strontium in seawater over the past 100 million years. The excess of strontium measured at the limit between the Cretaceous and Tertiary eras can be explained only by an extremely violent terrestrial impact of a meteorite 10 kilometers in diameter with ejection into the atmosphere of 10 to 100 times the mass of the meteor, or else by a meteorite 100 kilometers across.

The recent discovery of two extraterrestrial amino acids in Cretaceous/Tertiary sediments in Denmark also lends credibility to the hypothesis of a meteoritic impact. There is no reason to suspect sample contamination during analysis, because the two amino acids are quite uncommon on Earth. But how did the amino acids survive the collision with the Earth? The only explanation is that the bolide must have been very rich in organic matter.

There are other troubling questions about the impact theory. First, the impact must have left a huge crater, perhaps 200 kilometers in diameter, which has not yet been located. More recent and deeper measurements of the proportion of iridium in a certain number of sites do not tally very well with an instantaneous event on a geological scale. The excess of iridium spreads over more than 100,000 years in Italy but only 6,000 years in Canada. In Scotland, the rate remains high for 20 million years and there is no marked excess at the Cretaceous/Tertiary limit. A 24-meter-long core extracted from the bottom of the North Pacific shows an excess of iridium compatible with a normal flow of dust and meteorites. These observations are better explained by intense volcanic activity, such as that in the Deccan plateau, mentioned above. But perhaps repeated impacts could explain a longer period of iridium excess. A last criticism is based on the fact that on the Moon, constantly bombarded by meteorites, the excess of iridium is no more than 4 percent.

The proponents of meteorite impact and those of intense volcanic activity may both be right, if the former triggered the latter. Indeed, the impact could have produced a hole 20 to 40 kilometers deep — enough to cut into the Earth's mantle, thus causing an outpouring of lava that would have covered up the crater.

Whatever its origin, a large quantity of dust and soot in the Earth's atmosphere would have important consequences for the climate, and thus for life. It would block out the rays of the Sun, lowering the temperature on the ground and in the different layers of the atmosphere and leading in turn to important changes in the composition of the atmosphere as well as in the global climate. This prediction comes from extensive studies of the possible long-term consequences of a global nuclear war, which would also produce large amounts of dust and soot.

The massive extinction at the Cretaceous/Tertiary boundary is not unique in the history of our planet. The living world has in fact suffered many other massive extinctions, which call for a comprehensive interpretation and which may shed some light on the causes of this specific one.

The "normal" rate of species extinction in the living world is three to five families, i.e. 180 to 300 living species, per million years. During massive extinctions, this rate increases so dramatically that it is remarkable that any species survives at all. The most important extinction, dating from the end of the Permian era about 250 million years ago, saw the disappearance of 96 percent of marine invertebrate species in several million years. The second most important, and the oldest, was 440 million years ago. In 1983, David Raup and John Sepkoski, paleontologists at the University of Chicago, showed evidence for a periodicity of 26 million years in the extinctions of the past 250 million years. Among the twelve events they classified as "massive extinctions," five are unquestionable and saw the disappearance of up to twenty families each.

After the discovery of the iridium excess by the Alvarezes, this periodicity appeared as another important clue for understanding massive extinctions. Astronomers, geophysicists, paleontologists, and biologists all started looking for possible explanations of the periodicity, and the number of papers per year on massive extinctions jumped from 40 or 50 to over 500. Table 2 summarizes the suggestions. (A minus sign precedes causes which were later rejected.)

Table 2
Possible causes for periodic extinctions of the living world.

Astronomical causes	Terrestrial causes
Nemesis, companion of the Sun	Variations of sea levels
A tenth planet	Variations of global temperature
– Crossing of the galactic plane	Periodic tectonic activity
– Inversion of Earth's magnetic poles	Periodicity proper to the living world

Because of the considerable importance that most scientists ascribed to this periodicity, and because of the skepticism shown by some geologists, Raup and Sepkoski extended their analysis with new data in 1986. This second study confirmed that extinctions of families and especially of species during the past 250 million years indeed have a 26-million-year periodicity.

To explain these extinctions, astronomers invoked a periodic astronomical event which would indirectly provoke a bombardment of the Earth by asteroids, comets, meteorites, or a flow of energetic particles.

Nemesis is the name of a hypothetical star of faint magnitude which circles the Sun in 26 million years. Its very elongated orbit would cause the star to periodically cross "Oort's cloud," a reservoir of comets which is found 55,000 astronomical units[3] from the Sun. This disturbance in the cloud would send about a billion comets into the solar system, at a rate of about 66 per year. Twenty or thirty of them could collide with the Earth over a period of a million years or so. Unfortunately, such an orbit is very unstable, and Nemesis could escape from it in less than a billion years. It could also penetrate the solar system and create grave disturbances (for example, ejecting Jupiter) if its mass is greater than 0.05 solar masses. A group of astronomers in California has nevertheless begun to look for Nemesis among the 5,000 possible red dwarf stars by measuring their proper motion on photographs taken at intervals of 3 to 6 months.

The periodic swarms of comets could also be caused by a hypothetical tenth planet in our solar system, aptly called Planet X, located at an average distance of 50 to 100 astronomical units from the Sun. If its orbit is slightly elongated, the planet will periodically graze one of the edges of a comet disk whose existence we must also presuppose, causing a bombardment of comets. This tenth planet would also explain the perturbations observed in the motions of the outer planets, which Pluto's low mass cannot account for. The search for planets beyond Pluto in the solar system is on the program of the Hubble Space Telescope. This planet could also be discovered in the infrared range by the IRAS satellite (mentioned below).

The Sun oscillates periodically about the galactic plane and crosses it about every 30 or 33 million years. This periodicity suggests a third possible cause of the mass extinctions: the intense radiation of cosmic particles and the flow of x rays confined in the

3. An astronomical unit is the distance from the Earth to the Sun, 149.6 million kilometers. Pluto, the most distant planet, is about 40 astronomical units from the Sun.

plane. The quantity of interstellar matter involved would be much greater. Giant molecular clouds in the galactic plane could disturb Oort's cloud, indirectly provoking a shower of comets. But even though the period of the Sun's oscillation is very precise, this would not be true for these encounters because the clouds are spread over a certain thickness about the galactic plane. Furthermore, at present the Sun is almost in this plane, but we are approximately midway between two mass extinctions.

In the last astronomical explanation, suggested by Raup, Earth's magnetic poles would invert with a periodicity of 30 million years, and all the energetic particles from which we are usually shielded by the magnetic field would be able to bombard us. But a deeper statistical study of the dates of inversion of the magnetic field has not confirmed the presumed periodicity. It shows, rather, the difficulty of establishing periodicity from uncertain data.

Astronomers are keen to find astronomical explanations for the periodicity of massive extinctions, but plausible causes of these events may also be found on Earth. Geologists and biologists have proposed explanations of their own. Terrestrial phenomena, such as the periodic variations in sea levels, climatic variations (in particular, global temperature on the Earth's surface), or variations in the Earth's crust (tectonic movements), may also contribute to extinctions.

Abrupt climatic changes may occur in response to slow variations in the atmosphere and the oceans.[4] There are some good examples of rapid climate swings in records from the Quarternary era, usually involving glaciation, as well as in theoretical climate modeling by computer. The transition between a seasonally ice-free state and permanent snow cover, for example, occurs with a minute change in carbon dioxide. Although unprovoked and abrupt climatic changes are not likely to have caused the massive extinction at the Cretaceous/Tertiary boundary, they may be a viable explanation for other extinctions.

Two of the mass extinctions happened at the end of a period of stability in the Earth's magnetic field, and therefore of stability

4. On a geological time scale, "abrupt" means about 10,000 years.

in the Earth's core.[5] However, such stability would have to be compensated for by movements of the crust and heavy volcanic activity. But in principle these events have no reason to be periodic unless they are another consequence of the swarms of comets or meteorites suggested by astronomers.

Analysis of large craters on Earth seems to support astronomers' claims. About a hundred craters, whose estimated age can be up to 660 million years, are known. Statistical studies of 32 craters more than 10 kilometers in diameter, known to be less than 365 million years old, revealed a periodicity of 32 million years in their formation dates, or 28 million if the three latest are excluded.[6] Whatever the cause of the bombardment, it seems to have happened with approximately the right periodicity for extinctions. But the meteoritic matter found at the sites of eight craters is more characteristic of large meteorites and asteroids than of comets, and thus discredits theories of comet bombardment.

The correct explanation might again be a combination of several effects, because there also seems to be a periodicity in the major basalt volcanic eruptions in the past 250 million years, in phase with the massive extinctions. The triggering mechanism of these outflows is unknown, but it may be a periodic shower of extraterrestrial bodies, simultaneously explaining the periodic craterings, lava flows, and massive extinctions.

In all cases, the derived periodicity depends critically on the way the time scale was established by geologists. They consider that dates concerning the last 100 million years are trustworthy, but earlier dates can be uncertain by up to 5 million years. It is therefore not surprising that they would have proposed three competing time scales to date geological layers. For a given geological time scale, the age of different sediments can only be estimated within half a million years. Therefore we cannot be sure of the simultaneity of determined extinctions at different geological sites.

The way mass extinctions are identified — through the analysis of populations at successive geological stages — can also influence

5. The inversion of the magnetic poles is provoked by movements in the Earth's core.

6. One of them, Rochechouart Crater (France), 23 kilometers in diameter, is estimated to be 160,000 years old.

the observed periodicity. Using this method, massive extinctions cannot be found in two consecutive areas. This favors artificial periodicity. If, on the other hand, extinctions whose importance has been overestimated are included in the analysis, a random distribution of extinctions becomes more likely.

Astronomers were eager to find astronomical causes for a phenomenon for which the periodicity is not well established. Paleontologists must be given enough time to obtain more precise data on the sizes and the dates of the extinctions. The analysis of the proportion of iridium in sediments dating from the mass extinctions should confirm the hypothesis of extinction by impact. We already know that this proportion is abnormally high for three of these periods, although for two of them the excess has been found at only a single site.

The coincidence between the periodicity of certain astronomical phenomena, impacts, and extinctions suggests a causal relationship but proves nothing. Some biologists think that impact by meteorites would have much more catastrophic effects on the living world. The damage, always greater in tropical zones, would resemble that caused by El Niño.[7]

If astronomical or terrestrial events really do introduce periodicity in the evolution of life, the consequences at the conceptual level are important and deserve to be examined. Instead of a linear and irreversible evolution toward ever more complex organisms, we could imagine a cyclical one in which periodic cataclysms cause the living world to move backward. In such a case, each new start would produce new varieties of living species, according to circumstances. Cyclic evolution has been invoked in the past to interpret history, perhaps by analogy with the rhythms of our life. But such a scenario is not compatible with the irreversible evolution generally found in the universe, and more particularly on our planet.

We can also consider that extinctions have a stimulating effect on the evolution of life by freeing habitats for the development

7. El Niño is a meteorological phenomenon linked to modifications in ocean and atmospheric currents affecting both North and South America. Abnormally warm ocean currents and abundant rainfall in tropical zones have disastrous ecological effects. The last El Niño dates from 1982–83.

of new living species and thus diversifying the living world. We can rest assured, in any case; if extinctions are periodic, the last one has been set at precisely 11 million years ago by the three geological scales, and the next one is not due for another 15 million years.

Terrestrial and extraterrestrial catastrophes form an integral part of Earth's history. By discussing massive extinctions in the living world I have evoked catastrophes which could disturb, at regular intervals, the evolution of living species. Even though some of them are unlikely, they show how fragile life is.

Let me mention, for completeness, some other periodic events in the solar system which affect our climate, and which could become catastrophic under certain circumstances. The Sun's 11-year cycle of activity, during which sunspots appear and then disappear, has a well-known effect on radio communication but also on the climate. The small ice age of the seventeenth century coincides with Maunder's minimum, a period when solar activity was particularly calm. The higher average temperatures of the period from the eleventh century to the middle of the thirteenth corresponds to a more intense period of solar activity, when sunspots were visible to the naked eye. We can easily imagine that greater changes in solar activity would have catastrophic consequences for life on Earth.

The Earth's movement around the Sun undergoes small periodic variations which, if amplified, could also have dramatic consequences. The axis of rotation describes a cone around the vertical with respect to the plane of its orbit, with a period of 24,000 years. This is called precession of the axis of rotation. The inclination of this axis can also vary periodically; this *nutation* has a 42,000-year period. Finally, the Earth's orbit around the Sun is not a circle, and its eccentricity has a period of 102,000 years.

These periodic variations are responsible for the successive ice ages of the past 500,000 years and allow us to predict a new ice age in the Northern hemisphere several thousand years from now, unless man intervenes. These perturbations are very weak, and our Moon has a definite stabilizing effect. It is even possible that these perturbations have a beneficial effect on the evolution of life, because it must adapt to ever-changing conditions whereas a very stable environment might lead to a stagnation of biological evolution.

Other events, which are not periodic, can also disturb the evolution of life. Life on Earth depends critically on the Sun. But the Sun is not immortal. In about 5 billion years, hydrogen burning will be over, and the brief contraction that follows will start the helium burning phase and the expansion of the outer layers of the Sun, whose diameter will encompass Earth's orbit. By then, we will probably have attained a technological level that will enable us to emigrate to a more clement planet. It will, in any case, be the end of all evolved life on Earth.

This stage of evolution, known as the *red giant* stage, is reached much more rapidly in more massive stars because they evolve faster. Life around these massive stars will be endangered much earlier, perhaps even before it has acquired the means to emigrate from its traditional niche.

If a star close to the Sun (such as Sirius, which is only 8 light-years away) were to become a red giant a thousand times brighter than the Sun, the influx of additional radiation on the Earth's surface would be negligible. However, if Sirius became a supernova, the consequences for life on Earth would be catastrophic. One star out of a hundred ends its life in that spectacular way. For about 2 weeks it radiates away a billion times more energy than the Sun. The expanding shell of debris forms a cloud of hot gas and energetic particles which will emit radio waves and x rays for thousands of years.

How often is such a catastrophe likely to occur? The frequency of supernova explosions in a spiral arm of our Galaxy less than 30 light-years from the Sun is one every 100 million years. But this is a statistical estimate, and such an event, although rare, could happen any time. The flow of cosmic rays produced would reach the Earth in 100 to 1,000 years and would be 1,000 times more intense than the average flux received on the surface of our planet. Such a dose of radiation would reduce the transparency of the atmosphere in the visible range and increase it in the ultraviolet. The consequences for life would be disastrous: general cooling by

3°C, and destruction of living cells by ultraviolet radiation, and by x rays (which are deadly in the long term).

The passage of a star or a large asteroid in the neighborhood of the Earth could have catastrophic consquences, such as ejecting the Earth out of the solar system. In the solar system there are about a thousand asteroids more than a kilometer in diameter whose trajectory crosses that of the Earth. Table 3 shows the probable frequency of collisions. The last important collision we know of occurred at Tunguska, in Siberia, on June 30, 1908. (The colliding object may have been a comet instead of an asteroid. A New Zealand geophysicist, A. R. Crawford, even thinks that it was a catastrophe of terrestrial origin, due to a sudden outpouring of a large quantity of hot gases.)

Small meteorites fall to Earth very frequently. On a starry night we don't have to wait very long to see a shooting star. But these meteorites pose no danger to life, even to an individual. The only known occasion when a person was hurt by a meteorite was on November 30, 1954 in Alabama. In North America, of the sixteen meteorites found in the past 20 years, seven caused damage to buildings, usually to the roof. The annual frequency of meteorite showers has been estimated by a team of geophysicists in Ottawa. It decreases exponentially for large meteorites. The number of meteorites weighing 1 kilogram or more falling on the whole continental surface of the Earth has been estimated at 1,200 per year. We can say that only one person on Earth will probably be hit by a meteorite weighing a few grams or more in 9 years, and that sixteen buildings may be hit yearly.

Table 3
Probable frequencies of collisions between Earth and a meteorite (after Opik 1973).

Bolide's diameter (km)	Time between two collisions (million years)	Destroyed area (km^2)
2.1	13	320
4.2	62	840
8.5	260	2,200
17.0	1,100	5,000
34.0	4,500	11,000

I have inventoried astronomical phenomena — some recurring, some random — whose effects on the living world would always be catastrophic. Such events may have affected the date of man's appearance. They may even have influenced evolution in favor of mammals, just as they could have favored marsupials or insects. Man's appearance may be the result of chance, but I am convinced that, sooner or later, one species would have dominated the others — not necessarily through physical force, but perhaps through the greater complexity of its activity resulting from its superior intelligence.

This conviction comes from the history of the universe. In fact, the parallel becomes obvious if we place our planet's history in a wider context. Evolution toward an ever more complex organization is a universal phenomenon. If the evolution of life on Earth reflects that of the universe, it must inevitably lead to more complex and more intelligent organisms, despite any historical contingencies which may modify the course of this evolution.

4

Is Our Future in Space?

In Manchester the air is so polluted that one wakes up in the morning hearing the birds cough.

—Anonymous, *Le Monde*, August 6, 1982

In the preceding chapters I briefly discussed the origin and evolution of life on Earth. Even though this presentation does not give a definite answer to our questions, it nonetheless leads us logically to the threshold of one of our greatest problems: the future of our civilization.

On the front of the Centre Beaubourg in Paris, a clock has started counting backward toward the third millenium. The years to come will be crucial for humanity. At the dawn of the twenty-first century, a sense of emergency, disarray, and the menace of brutal mass destruction seem to dominate. The seconds which click away remind us that solutions must be found urgently.

The problem is easy to pose: Can a technological civilization such as ours hope to last a long time, or does its very nature doom it to disappear after a few centuries? This question does not concern only our endangered planet; it is valid for any life form which may have developed in a similar way. If the life span of a technological civilization is structurally short, it becomes difficult to imagine a heavily populated universe. If every birth and development of life, after a shorter or longer period of adaptation, leads to contradictions and destruction, then civilizations will follow one another in the universe like shooting stars in the sky, leaving only faint trails behind them.

There are many problems which can spiral into fatal and insurmountable crises. I will present those that astronomers and geophysicists are competent and duty-bound to analyze and evaluate, to inform their fellow citizens.

It seems evident that the principal danger currently threatening all of humanity is nuclear war. Its immediate effects would be disastrous, but those living beings which survived the blast and the radioactivity would then have to endure a "nuclear winter." The effects of such a winter have often (and hotly!) been debated by the scientific community since the beginning of the 1980s.

Carl Sagan is one of the authors of a global climatic model, published in 1983, that paints a gripping picture of the post-nuclear-war situation. The chief actors of the script are the hundreds of millions of tons of dust, smoke, and soot from the fires provoked by explosions. These clouds of particles and soot would rise with warm air currents and would cover the whole Earth permanently because of their low weight. They would reduce sunlight by at least 50 percent and bring on cold and darkness. Temperatures would not rise above −15°C for a year.[1] A second consequence is that the heat from the explosion and the nitrogen oxide produced by the fires would dissolve the stratospheric ozone layer,which protects us from the Sun's ultraviolet rays.[2]

The consequences of such a nuclear winter for humans would be disastrous: destruction of crops, production of toxic gases such as dioxin in chemical fires, increased exposure to harmful ultraviolet radiation. These effects would be compounded by the social consequences of war and destruction, including the breakdown of social structures and of basic social services, and the interruption of international trade (in food, energy, medicine, etc.), making nuclear war a major catastrophe on a global scale.

Since 1983, more sophisticated models have added more details, as well as nuances, to this apocalyptic picture. They confirm that dust and soot clouds would cover the planet in less than 2

1. By comparison, the 200 million tons of smoke now being emitted annually disturb the atmosphere only slightly.

2. The stratosphere is the atmospheric layer situated between 10 and 50 kilometers in altitude. Between the ground and the stratosphere is the troposphere.

weeks, but they suggest the nuclear winter would be less severe in January than in summertime because of the Sun's weak rays in the northern hemisphere in winter. In these models the explosive power used is 5,000 megatons — about half the present nuclear arsenal. The predictions of these models are in fact little sensitive to this amount, provided a certain threshold of explosive power has been passed. The climatic models have the same defects as weather predictions — they can diverge from reality very rapidly. For example, rains can be very efficient in washing the atmosphere and getting rid of dust. Let us hope that these predictions will have more effect on the superpowers than the peace movements do.

Atmospheric pollution, urgently in need of a solution, is another negative consequence of our unrestrained technological development. Through ignorance, indifference, or both, man is irreversibly altering the vital properties of the atmosphere. Two changes in the atmosphere with dramatic consequences for life have recently appeared as results of pollution: global warming and an increase in ultraviolet radiation from the Sun. Both of the changes are due to the release of gases produced by man.

Carbon dioxide emitted by fossil fuel combustion is one of the factors responsible for the global warming. With methane and ozone, CO_2 forms a thermal shield in the troposphere, about 10 kilometers up, which absorbs the thermal radiation re-emitted by the Earth's surface and thus maintains a mild temperature in the atmosphere. In the upper atmosphere and at present levels, these gases are indispensable to life, because without this greenhouse effect the temperature would be permanently at −20°C. But human activities keep increasing the proportion of these gases in the atmosphere. Carbon monoxide and carbon dioxide result from industrial and domestic combustion and automobile exhaust. Carbon dioxide in the lower atmosphere has increased by 25 percent in the past 200 years. Weekly measurements taken over 6 to 8 years at seven sites distributed around the globe (at the equator, in the tropics, at temperate latitudes, and at the poles) show that, on the average, carbon monoxide increases 0.8 to 1.4 percent annually.

Ozone also increases in the troposphere, but for different reasons. Nitrogen oxide of industrial origin combines with atmospheric oxygen to give off ozone. Two German chemists recently found a statistical bulletin of the city of Paris which published measurements of the quantity of ozone taken by the Observatoire de Montsouris from 1876 to 1910. Compared with the present rate, these measurements show that the proportion of ozone in the lower atmosphere has more than doubled in 100 years.

Furthermore, the chlorofluorocarbons, or CFCs,[3] although present in extremely small quantities (one part per billion) in the troposphere, also contribute significantly to the greenhouse effect. They have been shown capable of absorbing 10,000 times more heat than carbon dioxide. In 1981 the quantity of CFCs in the atmosphere was estimated at over 12 million tons. Even if their production were stopped today, their concentration in the atmosphere would continue to increase for another century. CFCs take 15 to 20 years to reach the upper atmosphere, where their life span is 60 to 100 years. Since CFCs do not dissolve in water, even rain cannot prevent their slow rise.

At the current rate of pollution, the average temperature of the atmosphere might increase by at least 1°C, or maybe even 5°C, by the year 2050. Land-based records indicate that it has already increased 0.5°C in the past hundred years.[4] The consequences of a further increase would be disastrous for human life: desertification of temperate zones, displacement of agriculture to northern Canada and Siberia, increase in sea levels of 1 to 5 meters, flooding of coastal regions, salination of drinking water sources, population migration.

The history of the Earth confirms the possible climatic effects of atmospheric carbon dioxide. Recently geophysicists discovered that there was a 40 to 50 percent increase in atmospheric carbon dioxide at the end of the last Ice Age, 10,000 years ago. They could not, however, decide whether there was a cause-and-effect

3. These gases, commonly termed *freons*, are used as aerosol sprays, as refrigerator coolants, as cleaning fluids, in foam rubber, etc.

4. This is a controversial result; data on ocean temperatures would be necessary to establish the reality of a global warming in the past century.

relationship between this increase and the warming. A 2,085-meter-long ice core extracted by a Franco-Soviet team from the Antarctic ice cap ended any doubts. The analysis of this core in 1987 showed that the proportion of carbon dioxide and the climate were intimately linked over the past 160,000 years. The proportion of carbon dioxide was high between ice ages, and low during them. The influence of sunlight, which varies with the Earth's orbit according to well known periods of 100,000, 40,000 and 20,000 years, was also shown. We thus have definite proof that a clear increase in the quantity of carbon dioxide in the atmosphere will have significant effects on the climate.

The second major casualty of atmospheric pollution is the stratospheric ozone layer. Each spring, this ozone layer thins dramatically above Antarctica. A team of researchers from the British Antarctic Survey revealed this fact to an amazed world in 1987. This ozone layer is so fragile and tenuous that, if it were brought to the temperature and pressure of sea level, it would only be 3 millimeters thick. In October 1987, above the Antarctic, this thickness was equivalent to only 1.3 millimeters at sea level.

Is this a temporary, localized anomaly of weather conditions specific to Antartica, or is it the first sign of irreversible destruction of the stratospheric ozone layer? At present, this layer protects us by absorbing 99 percent of the Sun's ultraviolet radiation. Its decrease, or even worse its disappearance, would be a catastrophe for life on Earth, as the above-mentioned increase in tropospheric ozone is much too small to make up for this loss. The incidence of cataracts and skin cancer would increase, plankton and many crustaceans in the oceans would be destroyed, and agricultural production would decrease.

Because of the urgency and the gravity of the situation, in the fall of 1987 NASA allocated $10 million to deal with it. One hundred and fifty researchers from nineteen organizations participated in the program, and two specialized airplanes — a DC8 and a modified U2 — flew over Antarctica for 6 weeks to measure the chemical composition of the upper atmosphere. The campaign concentrated on two possible causes of the phenomenon: winds (which might redistribute ozone) and CFCs (which destroy it).

The results showed that wind conditions indeed contributed to the damage. Although many uncertainties remain, unusual concentrations of chlorine compounds were also found, pointing to man-made causes.

It has been known since 1974 that CFCs could attack the ozone layer. The Sun's intense ultraviolet radiation in the stratosphere breaks up the CFCs' inert molecules into simpler but very active components. One molecule of these components can, all alone, destroy 100,000 molecules of ozone, if other reactions do not intervene. According to some estimates, a complete halt to the production of CFCs would not prevent them from reducing the ozone layer by 5 to 10 percent over the next 50 years. An international agreement was signed in Montreal on September 16, 1987, to curb the production of "substances which deplete the ozone layer" — mainly the CFCs, which presumably also contribute to global warming.

It is possible that the ozone hole over the Antarctic is simply a local phenomenon due to the fact that the atmosphere there contains less hydrogen and nitrogen than in other places, due to very special meteorological conditions, or due to winds that bring ozone-poor clouds from the troposphere into the stratosphere. According to an astronomer at the University of Kentucky, the quantity of ozone in the stratosphere varies with the cycle of solar activity, which just happens to have passed a minimum. Another culprit could be El Chichon, a volcano in Mexico, whose recent eruption disturbed the stratosphere. It is certain that we need data spread over a greater time period to separate seasonal or short-term effects from profound ones.

A recent study based on measurements taken by the polar orbiting satellite Nimbus 7 from 1978 to 1986 confirms our fears. According to this report, there is also an ozone hole over the North Pole, although it is smaller than the one over the Antarctic. Since 1983, large reductions in ozone are no longer confined to high latitudes, but also happen along the equator, where we find the lowest stratospheric temperature outside the Poles. Are all these changes found in the stratosphere cyclical or irreversible? We cannot yet see the event in sufficient perspective to measure its importance, but obviously it will have to be reevaluated frequently. This is precisely what most geophysicists involved in stratospheric studies worldwide are planning to do. NASA will renew its campain of 1987, this time over the North Pole, and in coordination with European teams. The Danes, the Norwegians, and the Finns will launch ozone probes and stratospheric balloons. The composition of the polar stratosphere will be monitored from Northern Scotland and the Spitzberg. The Soviet Union will also make a major contribution to these studies, in particular with a tropospheric and a stratospheric airplane.

We can easily imagine other crises. Overpopulation and ideological clashes come immediately to mind when we think of social problems. We are only beginning to be confronted with the potential dangers of genetic engineering, which allows us to create new bacteria through genetic manipulations on a microscopic scale. Some of these bacteria are useful for humanity, for example those which transform hydrocarbons or produce artificial hormones. But we could also imagine that such manipulations might be used for diabolical purposes, such as bacteriological weapons.

These are some of the internal crises that threaten our civilization. There will probably be others that we cannot foresee. They are all more imminent than the astrophysical catastrophes discussed in the preceding chapter. Paradoxically, they are due to the very development of our intelligence, to the rapid progress of our sciences and their technological applications. We modify our environment more and more profoundly, and we are constantly confronted with the unforeseen consequences of these changes, both ecological and social. These crises are so closely linked to the very process of elaborating our society that it is logical to think that they must be universal enough to be applied to any civilization whose technological level is similar to our own.

Even though the problems are posed easily enough, for those who do not have the gift of second sight it is hard to find solutions. Perhaps the way to the solutions lies in our becoming conscious of the reality of the danger. It is always possible that such crises are just the normal consequences of a too-rapid development of society, a sort of childhood disease for which a cure will be found, after which the life expectancy of our civilization will become very long — perhaps several million years. We can even speculate that, once these growing pains are over, many civilizations enjoy a harmonious evolution, communicating among themselves and waiting, perhaps, for us to join them.

With the extremely optimistic viewpoint that all the crises I have evoked will one day be resolved, in what form can we imagine the future of our civilization? More specifically, is our future in space?

The conquest of space is the logical culmination of all the progress that has taken place in this century. The stability of Western society is in large part based on this uninterrupted progress and on confidence in a better future. Even if space travel does not solve all our troubles, it will still be necessary, because our needs will soon outstrip the Earth's mineral and energy resources and our activities will undermine our fragile environment.

The Apollo program, with missions from July 1969 to December 1972, was the first important step in the manned exploration of the solar system. During this program, which cost $25 billion at the time, eighteen men were sent to the Moon and twelve actually walked on the Moon's surface. The number of successful missions, six out of seven, shows that NASA indeed mastered the technology of manned interplanetary flight. If this ambitious program was interrupted prematurely, it was for budgetary and not technical reasons.

The Soviet Union's space agency, Interkosmos, is acquiring important knowhow in preparation for future manned interplanetary flights, and now has a comfortable lead over NASA. In 1987, for example, 96 of the 111 successful space launches in the world were from the Soviet Union.[5] The Soviet cosmonauts began breaking records for endurance in space in the late 1970s with their second-generation space stations. The first crew of the Salyut 6 space station stayed in orbit for 96 days. Starting with Salyut 7, launched in 1982, stays in space became progressively longer. The present record is held by Vladimir Titov and Musa Manarov, with over a

5. Eight were from the United States, three from Japan, two from Europe, and two from China.

year on board Mir. This new space station, launched in February 1986, is a third-generation module. It is not only more comfortable for the astronauts, it is also better equipped technically, thus confirming the Soviets' intention of pursuing their efforts in space. The objective is a manned mission to Mars around the year 2020.

After a long period of crisis due to a lack of political and public support, NASA is now apparently regaining momentum. It has announced its intention of establishing a program for research and training in the space life sciences, and it began soliciting proposals from scientists in February of 1990. Its main goals are to ensure the health, well-being, and performance of humans in space, and to develop an understanding of the role of gravity in living systems. It is thus aimed at expanding human presence beyond low Earth orbits, beginning with the space station Freedom, to be assembled in space between 1995 and 1999. The other goals of this program are to expand our understanding of the origin, evolution, and distribution of life in the universe and to promote the application of this research to improve the quality of life on Earth.

It takes an average of 5 days to get to the Moon, and the Apollo missions usually lasted about 12 days. It takes 6 to 8 months to get to Mars, so sending astronauts there presents greater biological and medical problems than technological ones.

The Soviets are becoming very competent in space agriculture. They know how to grow plants in space that produce fertile seeds. These plants purify the air aboard a space ship by absorbing carbon dioxide, and also provide some of the astronauts' food.

The Soviets now also know how the human organism reacts to a prolonged stay in space, and in particular to the effects of weightlessness. One of these effects is an irreversible decalcification of bones. It is as though the skeleton were useless in the absence of gravity. Would the morphology of the octopus be best suited for this environment? But a rigid infrastructure is necessary so that muscles can produce movement. Among the other problems caused by weightlessness is space sickness, which affects about 70 percent of astronauts. Its first symptom is persistent nausea, when the mechanism in the inner ear which orients the body stops functioning. Finally, fluid distribution in the body changes, and blood tends to concentrate in the head and increase its volume. But the body adapts progressively to weightlessness; on their return to Earth, astronauts are usually incapable of getting out of their seats and need several days to get used to Earth's gravity again.[6]

A manned mission to Mars would be much less costly than the Apollo program in relative terms. In 1984, its cost was estimated at $34 billion (half the amount to get to Phobos, the other half to land on Mars). However, such a mission should not just be a brief stay to plant a flag and gather a few samples. It would be much more economical and scientifically profitable if planned as part of a vast program of exploring the solar system, with Phobos used as a starting point for missions to the asteroids.

A stopover on Phobos (or Deimos, Mars' other satellite) could be used for refueling. This bold idea for saving weight at takeoff has its supporters at NASA as well as at Interkosmos. Phobos probably contains water. This water, decomposed through electrolysis, would provide oxygen and hydrogen — whose explosive mixture serves as rocket fuel. Because Phobos has only a thousandth of Earth's gravity, a space vessel would have no problem escaping.

All of the concrete achievements in space and the projects which have arrived at advanced stages suggest that man is preparing to conquer space. But before we go on to discuss the even more futuristic plans for space colonies, we must determine the present limits of our capacity to travel into space. The recent crisis in space transportation has shown that the objective will not be reached in the near future.

Let me summarize the successive disasters which marked the year 1986. On January 28, the space shuttle Challenger explodes 72 seconds after takeoff. On April 18, a Titan 34D rocket explodes 8 seconds after taking off from Vandenburg Air Base, damaging the launch pad. On May 3, a Delta rocket takes off from Cape Kennedy; 71 seconds later one of the motors fails and the booster is destroyed. On May 30, an Ariane rocket lifts off from Kourou;

6. To speed up the readaptation process, they spend half an hour daily in a tank which creates a vacuum in the lower part of the body, thereby attracting the blood to it.

4 minutes and 36 seconds later, the third-stage rocket doesn't ignite and the rocket falls into the ocean. This run of bad luck should come as no surprise. In the climate of competition and haste which then reigned in the various space agencies, the order "launch when sure" progressively became a more risky "launch at any price."

This predicament reveals the growing difference between the needs of the military and industry, on one hand, and the long-term ambitions of NASA, on the other. The latter wants to maintain and develop a human presence in space, and is imitated by the European Hermes program and the USSR (which has also created a prototype space shuttle as well as a small space plane). With this strategy, the optimization of space techniques currently needed by the military and industry, who do not need men in space, is sacrificed to a distant future. The error committed was involving human operators in two functions which could be automatic: sending heavy payloads into orbit and recovering them. Human presence on board requires maximum security measures, which are compensated by savings brought by recycling the vehicle and the payloads.

The various national space agencies have short-term projects for establishing permanently inhabited space stations around Earth, but some of them also have plans for colonizing space on a larger scale in the distant future. The physical and climatic conditions necessary for life are very special and will probably not be found on the other planets of the solar system. It thus appeared simpler to build colonies from scratch in interplanetary space, instead of modifying the environment on another planet to accommodate human needs.

Artificial colonies in space must reproduce the characteristics of the earthly environment, especially the force of gravity. For that, they must spin. NASA had at first considered wheel-shaped colonies, but decided that pairs of cylinders would be cheaper to build and easier to protect from cosmic clouds and ultraviolet radiation. A cylinder about 10 kilometers long must spin very rapidly to simulate Earth's gravity — once every few minutes.

But this rotation is too rapid to provide the alternation between day and night.

(NASA)

A project for pairs of cylindrical colonies. The giant petal-shaped shutters collect sunlight which is then reflected into the cylinders through transparent walls. Periodically, the shutters fold onto the body of the cylinder to simulate nocturnal darkness.

In the plans for a cylindrical colony proposed by Gerard O'Neill of Princeton University, natural sunlight is sent into the interior of each cylinder by three aluminum mirrors through three transparent walls along the cylinder. The alternation of day and night and of the seasons is accomplished by maneuvering the mirrors, closing them over the "windows" at night. The two cylinders can be in opposite phases like the Earth's two hemispheres, with summer in one corresponding to winter in the other. The length of day determines the average temperature and the season. Variations of about 1°C must be reckoned. One of the ends of the cylinder is always turned toward the Sun, and a solar reactor provides the necessary 120 kilowatts of energy per inhabitant. The two cylinders are 90 kilometers apart, and each one is surrounded by a ring of 72 small agricultural zones, 35 kilometers away from the cylinders. In these zones, the atmosphere is sterile;thus, there is no need for insecticides. The climate in each zone is adapted to a certain kind of culture. Industrial activity also takes place in sites located away from the two cylinders, but under weightless conditions.

There are definite advantages to setting up earthly industrial activities on satellites around these colonies. First of all, Earth's environment would be rid of most industrial pollution. The solar wind would evacuate industrial waste into the interstellar void. The reduced gravity and the vacuum present real advantages, as recent experiments on the space shuttle have shown. The principal industries concerned are metallurgy (perfect alloys), production of ultra-pure glass for lasers and crystals for the electronics industry, and biological procedures (serums and vaccines).

Table 4
Characteristics of cylindrical colonies.

Length (km)	Radius (m)	Period of rotation (sec)	Population of two cylinders
1.0	100	21	10,000
3.2	320	36	150,000
10.0	1000	63	1,000,000
32.0	3200	114	10,000,000

Space production of energy is necessary because of the limited energy resources on Earth. Furthermore, it would considerably ease interplanetary colonization. The Sun is the energy source of the future. By the time it reaches the Earth, solar energy is already diminished, providing 12,200 kilowatts-hours per square meter per year. A solar power station needs an area of 820 square kilometers, a circle with a radius of 16 km, in order to produce a trillion kWh/yr with 10 percent efficiency. Mercury's soil could provide the materials for huge power stations that would orbit the Sun, sending solar energy to Earth and its colonies by means of giant reflectors.

(NASA)

The interior of a cylinder for 10,000 inhabitants. Tomorrow's habitat is inspired by today's models on Earth; watercourses are set up in urban zones, green areas have both an ecological and an aesthetic purpose. Farming is done in satellites at the ends of the cylinders.

The construction of the smallest colony of table 4 would take about 500,000 tons of material, 98 percent of which can be extracted from the Moon's soil. Its composition was determined by the Apollo missions: 12 percent aluminum, 15 percent iron, 40 percent oxides, traces of titanium. The oxygen needed for the 50,000 tons of water for the colony can be obtained from the oxides. The 5.5 tons of liquid hydrogen would be transported from the Earth. Minerals would be mined on the sites of the colonies. Raw lunar rocks would be sent to the site in parcels of 20 kg by a magnetic catapult or by a 4-km linear accelerator.

(NASA)

A solar eclipse seen inside a colony. The Earth's shadow advances down the valley. City lights can be seen in the background.

Since the Moon is smaller than the Earth, the speed necessary to escape the Moon's gravitational field is only 2.4 km/sec, versus 11 km/sec on Earth. The energy necessary is only 5 percent of what would be needed on Earth. Later, perhaps mineral resources could be extracted from asteroids — especially those resources rare on the Moon, such as carbon and metals in the platinum group.

Meteorites do not present much danger for space colonies. Around Earth, most meteorites are of cometary origin, probably made of a combination of dust and ice, instead of rocky fragments of asteroids. One collision with a meteorite weighing 100 grams is likely once every 3 years; for a large meteorite of 1 ton, the frequency is one per million years. On the other hand, the effects of microwave radiation on the atmosphere of a colony, and its ecological consequences are not well known and must be studied.

The site considered for the first space colonies is near the equilibrium point L5 in the force field of the Earth and the Moon, about 400,000 kilometers from both of them.[7] A quasi-elliptical orbit about 80,000 kilometers around L5 is stable.

Artificial colonies such as those suggested here would reduce societies to small autonomous entities the size of a city. This may be an ideal size; among the most creative human societies were Athens in the fifth century B.C., Florence and Venice in the fourteenth and fifteenth centuries A.D., and the Hanseatic cities a bit later.

I am not implying that such space colonies would definitely solve all the crises which threaten our civilization, except perhaps for pollution. But by increasing everyone's living space, they would temporarily relieve overpopulation and the social tensions which arise from it. All other problems of our civilization will have to find solutions that rely not on technological progress, but on increased social awareness and responsibility and on education.

If artificial colonies in interplanetary space are still science fiction, colonizing the Galaxy must seem all the more utopian. But this idea should be examined because of its possible generalization. If it is possible that we will spread our living space far beyond

7. In the force field of two bodies rotating about one another, there are five positions, called Lagrangian points of equilibrium, where the gravity resulting from the two bodies is nil. A satellite does not have to use any energy to stay indefinitely at one of these points.

the solar system in several hundred or several thousand years and then occupy part of the Galaxy, other intelligent societies — if they exist — may already have done so.

Any idea of crossing the astronomical distances which separate us from the stars by our present means of transportation is ridiculous. Interstellar travel at relativistic[8] speed, inconceivable at the present time, is indispensable for reducing the length of space voyages to a human scale. A hypothetical probe traveling at the speed of light would take 100,000 years to cross the Galaxy. But we must not forget that the measurement of time is relative. These years are measured by a clock at rest on Earth. For astronauts on board a probe launched under optimal conditions, the trip lasts less than a generation, as we see in table 5. These conditions are a constant acceleration equal to one g till mid-journey, then a constant deceleration until the end. One g is the acceleration (i.e. the force of gravity) at the Earth's surface.

Table 5
Relative durations of interstellar travels for a vehicle at constant acceleration of 1 g to halfway, and equal deceleration to the goal.

Duration of round trip (years)		Distance reached
on board	on Earth	(light-years)
5	6.5	1.7
10	24	9.8
20	270	137
30	3,000	1,450
40	36,000	17,600
50	420,000	208,000
60	5,000,000	2,400,000

The technical problem is simple to formulate: for a probe to reach 99.9 percent of the speed of light, it needs 2,000 times its mass

8. The term *relativistic* means that the situation depends not on Newtonian mechanics but on Einstein's relativity theory — either because the masses in question are very large or because the speed is close to that of light. (The latter is the case here.)

in fuel (assuming complete and 100 percent efficient conversion of mass into energy).

An ingenious solution is to use the interstellar medium as a source of energy and as a carrying fluid in a kind of nuclear reactor motor. In this case, the voyages have to take place in regions where the interstellar medium is dense enough. The problem is to direct the interstellar medium toward the motor, and at the same time to protect the vessel and the passengers from the induced flux of cosmic radiation. But this important difficulty is not the only one. We must also find a way to parry the other dangers of interstellar space: collisions with meteorites, passage near black holes, etc.

Given the speed of technological evolution, it is perfectly feasible that these technical problems will be resolved in a few hundred years. If we need any convincing, all we have to do is look at the rapid progress made in air transportation since the beginning of the twentieth century. The present exorbitant cost of manned spacecraft can only decrease as the number of passengers increases. We already have a reusable vehicle — the space shuttle. Now we have to increase its reliability, its efficiency, and its size, and modify the propulsion system.

The obstacles to interstellar colonization are probably not technological but human. The reasons for which we hesitate to consider the colonization of the Galaxy inevitable are connected with the "why" of such an undertaking. Recently astronomers questioned NASA's long-term project of putting a permanent human presence in space. Such a reaction, called the "van Allen syndrome" after one of their spokesmen, may appear surprising. It is, however, understandable if we count all the scientific space experiments which were sacrificed over 10 years for the space shuttle. Without being affected by this syndrome, we can try to examine critically the motivations that could, in the long term, lead to a significant human presence in space.

An internal crisis, of the kind discussed earlier, or an earthly or astronomical catastrophe could force emigration from Earth. However, such an event would not necessarily push us to colonize the whole Galaxy. All we would need to do would be to arrange our habitat on a neighboring planet.

But in about 5 billion years the Sun will become a red giant. Its diameter will encompass the Earth's orbit. At that moment, if humans still exist, they will certainly have the necessary technology to move their habitat, or build another one, about as far away as Jupiter. A few hundred million years later, the Sun will become a white dwarf with a much smaller diameter and very faint luminosity. Mankind would then move again, this time back into the inner solar system. That move would probably be the last, because the white dwarf stage can last tens of billions of years. Only the explosion of a supernova in the neighborhood of the solar system could force us to move hundreds of light-years away to escape the violent radiation, which would last tens of thousands of years.

A net increase in population leads to a decrease in everyone's living space and an increase in social conflicts. In order to maintain our living space, we would have to send 200,000 people into space daily. Even a constant rate of growth means that population is increasing ever more rapidly. Civilization must then expand into interstellar space more and more rapidly in order for everyone's living space to remain constant. But the speed of expansion cannot pass the speed of light. When this unsurpassable limit is reached, individual living space begins to decrease, even if colonial expansion continues. This limit sooner or later forces any civilization to control its birth rate. Thus, colonization of the Galaxy would alleviate overpopulation problems only temporarily.

Table 6
Space colonized when the expansion velocity of the colonizing civilization reaches the speed of light. From that date and distance, the living space of each individual can only decrease, because the expansion velocity can no longer increase.

Radius of colonized space (light-years)	Rate of growth (percent)
490	2.0
978	1.0
1,956	0.5
4,890	0.2
9,780	0.1

There are of course other reasons — social, political, religious, economic — for migrating. Curiosity and the spirit of adventure are further motivations. But there are also many human obstacles to colonizing the Galaxy. The considerable distances between stars mean very long interstellar journeys. Unless travel is at a speed close to that of light, trips could last several generations. At the end of the journey the colonizers would not be sure of finding surroundings perfectly adapted to their physiological needs. They would probably have to live in an artificial environment. Finally, the constraints of such trips (life in a limited space, reduced rate of growth) would be precisely those that some colonizers were hoping to leave. The advantages of colonization may thus appear too distant to motivate such expeditions.

We can draw an interesting parallel between colonizing the stars in our Galaxy and colonizing the islands of the Pacific Ocean. Ben Finney, an anthropologist at the University of Hawaii, considers that the balance has not always been positive in the colonization of the Pacific, even on large, fertile islands. In many cases, demographic growth rapidly exhausted an island's resources and created social tensions, as in Tahiti or in the Hawaiian archipelago. Easter Island, where by destroying the forests the colonists condemned themselves to stay because they could not build boats, is an extreme example. The fact that Polynesians did not colonize the American west coast, or at least left no trace of an attempt, also merits consideration. Was it lack of motivation, insufficient technology, or a hostile environment? Thus, although man has a colonizing spirit and innovative qualities, the history of failed colonization and shaky motivation shows that colonizing the whole Galaxy is not necessarily inevitable.

The scientific and technical progress on which we earlier based our hopes for interstellar travel could, on the contrary, make us lose all motivation for leaving a planet which could become a pleasant place to live thanks to them. Elimination of biological aging through medical progress could lead people to avoid possible accidental death in risky interstellar voyages. There are always some brave souls ready to go off on adventures, but if they are only a minority there may not be enough for colonization.

We must also remember that leaving our planet is not the same as the emigration of colonists in past centuries. A picture that remains in my mind is that of astronauts floating in space, but attached to the space capsule by a long umbilical cord. This cord is a very strong symbol of our attachment to our Mother Earth. Will we ever be capable of cutting it?

The benefits of galactic colonization now seem outweighed by the constraints imposed by interstellar travel and life in an artificial environment, as well as by the hopes of a more inviting future on Earth. But a bold extrapolation of our technological, social, and physiological evolution may convince us that the future is indeed in space. Going around the world in less than 24 hours was unimaginable a few centuries ago; now it seems banal. Our descendants will probably feel the same way about a trip to the star Proxima Centauri. Interstellar voyages can then be seen as a dynamic process rather than a restrictive one, a natural activity for highly evolved beings accustomed for generations to living weightless in interplanetary and then interstellar space. Eight million years separate us from *Homo faber*, but probably only a few thousand years separate us from *Homo galacticus*.

Life in the Universe

5

The Search for Life in the Neighborhood of the Sun

The well questions the night
so that a stone will fall in.
The leaf questions the sky
so that a bird will fall.
The whole sky questions man
and man questions the well.
—André Frédérique, *White Stories*

The first place for us to search for life in the universe is, of course, our own solar system. However, despite all that has been written about Selenites, Martians, Venusians, and Saturnians, the blue planet — Earth — seems to be the only one to harbor evolved life. No geometric structure has ever been detected on the surface of another planet. No exceptional radio signal has ever been received by our radio telescopes, other than the radio bursts produced by the satellite Io when it enters Jupiter's magnetosphere. We must therefore look for *elementary* forms of life in the solar system. We shall see that, although most of the planets offer little hope, some of their natural satellites could contain some life forms similar to those on the primitive Earth.

But before exploring the solar system's planets and their satellites, we have to define what we are looking for — that object of infinite complexity to which we give the simple name *life*.

By life, I mean life similar to that we already know on Earth. It can be defined by its properties: metabolism, reproduction, adaptation, selection, and evolution. Furthermore, life on Earth cannot function without a permanent source of energy, in this case the Sun. Some of its properties can be found in inanimate systems, such as the metabolism of fire or the reproduction of crystals. But these are quite simple systems, whereas life is characterized by a high level of organized complexity. This last criterion can give us a more general definition of life: a complex, organized, and open system. The structure of a city satisfies this definition, even though it is just a manifestation of life. But this slight difference of meaning is unimportant for us when we are looking for all possible indications of life in the universe. Living organisms can also be defined by their finality; in other words, there is a purpose to their existence. According to Jacques Monod, "in their structure and performance, they achieve and continue a project." This project is conserving and reproducing the norm.

But among all these definitions, which ones are useful for an astronomer? He will keep in mind the term *metabolism* (which implies transformation, and therefore residues[1]) and the term *complex* (which suggests the necessity of very diverse chemical reactions). Such signs are detectable by an exhaustive analysis of the light reflected by the planets in the solar system.

Life on Earth is based on the chemistry of carbon and water. Could we imagine a kind of life based on the chemistry of another element? Probably not. As a solvent, water has no serious competitors; ammonia liquefies at low temperature and cannot coexist with oxygen. Among the more abundant elements in the universe, only carbon and silicon can form long chains and thus the complex molecules necessary for life.

Only carbon permits such a large variety of chemical reactions. Despite its capacity to form long molecular chains, silicon seems unsuitable for life for several reasons. Silicon does not react easily. Unlike carbon, it does not have double and triple bonds, which allow biochemical reactions. Silicon and its compounds liquify only at very high temperatures. Carbon chains are twice as strong as silicon chains. Carbon dioxide dissolves in water, where organic chemistry developed. Its equivalent in silicon chemistry, silicon dioxide (sand), does not. Silane, the equivalent of methane, is very unstable. And finally, to all these reasons we can add one of common sense: if life could reasonably be based on silicon, such

1. In the case of life on Earth, these residues can be methane or carbon dioxide.

life would have a good chance of existing on Earth, where silicon is 600 times more abundant than carbon.

One could also imagine life based on complex crystalline, rather than molecular, structures. Jean Schneider of the Paris Observatory has suggested that line dislocations[2] in crystals could be the starting point for complex networks eventually giving rise to life. However, he has not explored all the consequences of crystalline physiology, which thus remains very hypothetical.

The presence of carbon, oxygen, and organic molecules in stars and the interstellar medium, and the organic chemistry that takes place in molecular clouds, are the other elements that demonstrate the universality of carbon chemistry and favor life based on organic molecules, even if we cannot categorically rule out other alternatives for the chemistry of life.

There are also organic molecules in other galaxies. A French radio astronomer, Leonid Weliachew, discovered the first extragalactic molecule about 15 years ago. Carbon monoxide has now been found in about fifty spiral galaxies. Its detailed mapping is very useful because it allows us to determine the bulk motions within these galaxies. More complex organic molecules, such as ammonia and formaldehyde (containing four atoms), have been discovered in about ten galaxies. In 1987, the new Franco-German millimetric interferometer allowed Fabienne Casoli and Françoise Combes of the Ecole Normale Supérieure de Paris to discover a giant molecular complex in a spiral arm of our neighboring galaxy, Andromeda. Organic chemistry is therefore not restricted to the Milky Way. If life exists elsewhere than on Earth, it could well be spread throughout the whole universe.

We must, nonetheless, enlarge our concept of life. Even though some prior notions seem indispensable – source of heat, atmosphere – we shouldn't be too "chauvinistic." Oxygen, which enables us to survive, is not indispensable for all life forms, and is corrosive and toxic in strong concentrations. Some living organisms survive in extreme conditions of temperature and pressure. In 1983, bacteria were discovered living at a temperature of 250°C

2. Dislocations are imperfections in the regular structure of crystals.

and under a pressure of 265 atmospheres. There could be living beings insensitive to ultraviolet radiation, perhaps thanks to a thick shell like a turtle's.

The only extraterrestrial site where we are sure there is no life is the Moon. NASA's Apollo missions 11 through 16 brought back 382 kilograms of lunar rocks, which were analyzed in the laboratory. An average of 15 percent of the rocks' mass is carbon. Although the techniques used in the analyses would have found infinitesimally small quantities of organic compounds (on the order of a nanogram), they found no amino acids, no fat compounds, and no other organic molecules. Most likely, any multiple carbon-carbon links would have been destroyed by the Sun's ultraviolet radiation and by micrometeorite bombardment on the Moon's surface.

Photographs of Mercury taken by Mariner 10 in 1974 showed a lunar-type surface covered with craters and wrinkles. Because of its proximity to the Sun, Mercury has no atmosphere and the flux of energy it receives is 10 times as strong as on the Moon. Finally, Mercury makes one rotation in 2 months. It is therefore no more probable that we would find life on Mercury than on the Moon.

Venus has a very dense and corrosive atmosphere, mainly composed of carbon dioxide and nitrogen and also containing water vapor and hydrochloric acid. The planet is permanently covered with clouds, which cause a greenhouse effect that makes the temperatures there very high — about 500°C. The pressure at the surface is 90 times higher than on Earth. Proteins are too fragile to survive in such an environment. The images of the surface sent back by Soviet probes show it to be geologically quite young, but the presence of stones suggests erosion. This may be aeolian (wind) erosion, despite the enormous pressure. The planet's period of rotation has been estimated from radar measurements to be 243 Earth days.

The first time we look at Mars through a small telescope, we are disappointed to see a fuzzy disk, with no trace of relief, and especially no canals. To see details on Mars' surface, we have to go to a site located at a high altitude, which favors good resolution, or else wait for a night when the sky is very transparent, as sometimes happens in winter.

The first detailed observations of Mars' surface were made in 1877 by the Italian astronomer Giovanni Schiaparelli. He called the long dark furrows that streak the planet's surface *canali*, but drew no conclusions about their nature.

At the beginning of the twentieth century, the American astronomer Percival Lowell popularized the idea that the *canali* of Mars were real canals, carrying water from the polar caps toward the equatorial deserts. The implication, of course, was that Mars was inhabited. Lowell based a whole theory on this hypothesis and started a debate which lasted several decades. The drawings made by astronomers looking through their telescopes certainly looked like very geometrical and suggested intelligent intervention. But at which end of the telescope was the intelligence?

In fact, the canals of Mars are caused by the way the eye perceives reality. It tries to put order into what it sees by recognizing the outlines and, in the case of Mars, by joining a series of dark spots into geometric patterns. The idea of canals on Mars would never have been considered if at the beginning of the century there had been sufficiently sensitive photographic emulsions. The 7,232 images of the Martian surface sent back by the American space probe Mariner 9 indeed show the existence of valleys, such as the 5,000-kilometer-long Coprates, but these do not appear to be artificial. The existence of permanent rivers in the distant past seems unlikely. The furrowed surface that is now visible was probably produced by sudden floods when the polar ice caps melted.

The "canals" were not the only Martian phenomenon to excite the interest of astronomers at the beginning of the century. There was also the seasonal change of colors on the surface. In this case too, optical illusions caused misinterpretation. However, seasonal variations of the size of the polar caps were a definite fact.

Although it is further away from us than Venus, Mars is the first (and still the only) planet where we have tried to find life by on-site biological experiments. The two Viking probes launched from Cape Kennedy in the summer of 1975 landed in July and September of the following year on two Martian sites 2,000 kilometers apart. Their identical equipment included photographic cameras and devices to analyze the soil and the atmosphere and to perform biological experiments. Both the probes and their equipment had been carefully sterilized before being launched so that they would not be detecting microorganisms brought from Earth.

(NASA)

The surface of Mars. The Martian landscape in the area called Utopia, where the Viking 2 probe landed, looks rocky and desert-like. The ground is covered with fine, red dust.

The photographs transmitted from the probes are now well known. They show ground which is red, stony, and deserted, with no trace of vegetation or any other life on a human scale. Of course, a life form unknown on Earth or of microscopic size could escape photographic detection.

Analysis of the soil's composition was disappointing too. A sample placed in an oven and heated in stages until it reached 500°C gave off carbon dioxide and a bit of water vapor. If organic particles had existed on Mars, even in the proportion of one part per billion, they would have been detected with this method. There were not even any organic compounds of meteoritic origin.

Their absence proves that they are rapidly destroyed by the Sun's ultraviolet radiation.

The two Viking probes analyzed the Martian atmosphere during their descent and then on the ground. Near the ground the atmosphere is composed of carbon dioxide (95 percent), nitrogen (2.5 percent), argon (1.5 percent), and other gases (in very small proportions). The pressure is about 135 times less than on Earth. The absence of water in the soil and in the atmosphere is a serious obstacle to any form of terrestrial life.

Three biological experiments were intended to reveal any microscopic living organisms. These were started 8 days after the first probe landed. To everyone's surprise, the results were positive.

In the first experiment, called "gas exchange," a nutritious solution was added to a sample of Martian soil, and any changes in the atmosphere that might be produced by microbian metabolism were registered. Significant releases of oxygen as well as carbon dioxide were detected, but the speed and the short duration of the reaction suggest that it was a purely chemical reaction and not an organic one.

In the second experiment, "release of marked gas," the food was radioactive. A radioactive gas was released, but this gas (certainly carbon dioxide) was probably the result of chemical reactions with peroxides.

"Pyrolysis," the third experiment, was an attempt to synthesize organic matter in the Martian soil from radioactive carbon dioxide and monoxide. The two gases exist in the Martian atmosphere, but they are not radioactive. After 5 days of incubation, the soil sample was analyzed to try to detect radioactive organic matter. It was positive in seven tests out of nine, but the scientists responsible for the experiment remained skeptical. One of the positive samples had been heated to 175°C for 3 hours before the experiment; no earthly organism would have survived such treatment.

Biological experiments have revealed that the Martian soil is chemically very active; some of the reactions registered could be of biological nature. But after several years of laboratory experiments on Martian-type soils, scientists are convinced that they have not found life on Mars.

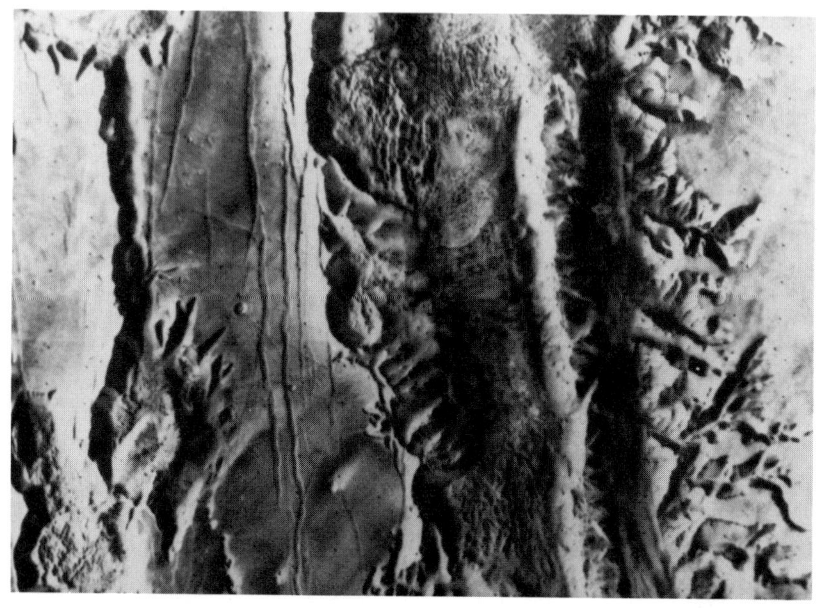

(NASA)

The surface of Mars: *Valles Marineris.* This image of the large canyon obtained by Viking 1 is a mosaic produced from 15 photographs taken with different filters to reproduce colors.

This does not definitely exclude the possibility of some form of life on Mars. The two Viking probes could have unluckily landed in desert regions. There is still also the possibility of finding life by digging deeper into the soil. Indeed, Martian organisms could have burrowed deep into the ground to protect themselves from meteorites and solar radiation, which the very thin atmosphere does not stop. And the Viking probes had only a very limited array of life detectors. There were no microphones, in case one of the Martian rocks had told the probe that it was illegally parked. A more realistic possibility is that the probes arrived too late to discover living organisms.

The lack of strong evidence for life on Mars has prompted suggestions that we look for evidence of *past* Martian life — in other words, fossils. If the water that produced the riverbeds on Mars were present long enough (say 500 million to a billion years) in the past, chances are that we would discover stromatolites and other fossils in the sediments. The main problem is that we have at present no way of dating the Martian surface. The same suggestion was made in the 1960s concerning the Moon, but the "sediments" in the lunar seas turned out to be basalts of volcanic origin.

The Viking probes did acquire images of strange features of the Martian surface, which made the cover of the scientific journal *Applied Optics* in 1988. They are located in the northern hemisphere, in the region of Cydonia Mensa bordering Acidalia Planitia and the northern plains. The main feature is less than 2 kilometers across and resembles a face; image treatment to remove shadows enhances this impression. In addition, there are a series of geometrical features 15 kilometers to the southwest: several pyramids and an unusual trapezoidal object resembling a fortress, all about 2 kilometers in size. Although I was a bit shaken when I first saw these images, I think the anthropomorphic interpretation of the features is just wishful thinking. The next missions to Mars should settle the point.

In future space missions, priority will be given to direct geochemical analysis of Martian soil, with, if necessary, a return of samples to Earth. The Soviet mission Phobos, made up of two identical probes carrying several Western experiments, was launched toward Mars and its satellite Phobos on July 7 and 12, 1988. Unfortunately, contact with one of the probes was lost in September 1988. After 10 months of travel in the interplanetary medium spent studying the Sun and the solar wind, the second probe took up an eccentric orbit around Mars for a week to study the planet's magnetic field and its interaction with the solar wind. Bad luck struck again when contact with this second probe was lost on March 28, 1989. The probe nevertheless returned a valuable harvest of new and unexpected results, including the discovery of a 1,000-meter-high cliff near the equator and the detection of carbon monoxide in the atmosphere. However, no ozone layer and no organic molecules were found, further reducing our chances of ever discovering life on Mars.

The Soviet project of exploring the red planet, of which the Phobos probes were the first step, foresees sending an automatic interplanetary complex toward Mars in 1994. It would include an orbital station, a vehicle, a balloon probe, surface probes and penetrators, and a small meteorological station. The "Marsokhod" vehicle would move over several tens of kilometers and take soil samples several meters deep, whereas the Viking probes could only scratch the surface. Another mission bringing samples back to Earth is proposed before the year 2000.

NASA, for its part, will send a Mars Observer in 1992 to make climatic and geochemical studies in the planet's atmosphere and on its surface. The recent discovery of iron carbonyl in Mars' atmosphere perplexed planetologists; on Earth, lead carbonyl can be produced only by a living process.

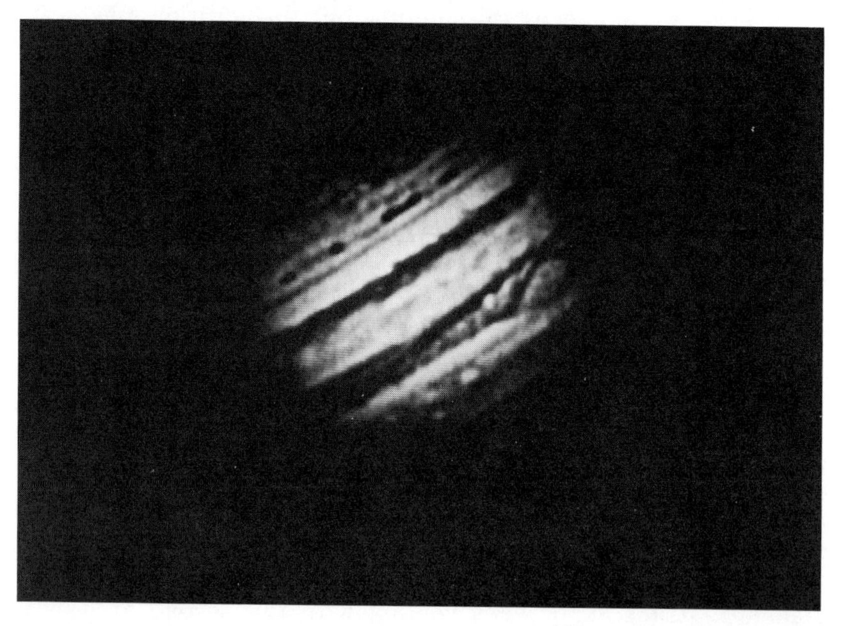

(Observatoire du Pic du Midi)

Jupiter. This is one of the best images of the planet Jupiter ever taken from Earth. It was taken in 1987 with a CCD camera at the one-meter telescope at Pic du Midi Observatory (France) by Eric Thouvenot and Christian Buil. Details of the turbulent atmosphere and the red spot can be clearly seen.

Jupiter and Saturn appear more like failed stars than planets. In fact, these two planets are almost 5 times less dense than Earth.

They are thus essentially liquid, probably formed of hydrogen and helium, elements which are also the main constituents of the Sun. They even contain a weak internal source of heat, probably due their slow contraction. Only their low mass, insufficient to start a catastrophic contraction, differentiates them from stars. The atmospheres of the two giant planets are rich in hydrocarbons and probably also in nitriles, despite the presence of molecular hydrogen. However, laboratory attempts to produce organic molecules as simple as formaldehyde in an environment similar to their atmospheres have failed.

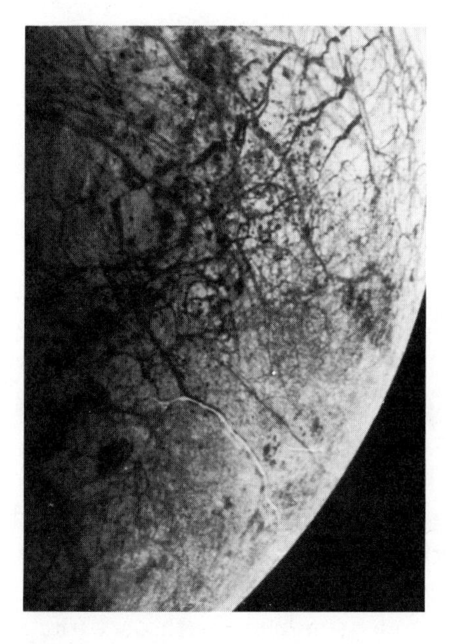

Europa. The surface of this satellite of Jupiter is streaked with cracks about 10 kilometers wide and several thousand kilometers long. This icy crust may cover oceans of liquid water and primitive life.

(NASA)

The Voyager probes that explored Jupiter, Saturn, and their satellites in 1979 and 1980 revealed worlds of unsuspected variety. One of Jupiter's Galilean satellites — Europa — is covered with oceans of liquid water under a thick layer of ice. Since life on Earth started in the oceans, Europa (which is slightly larger than our Moon) deserves our attention. The physical environment (temperature and pressure) is suitable and stable, as the presence of water in liquid form attests. Except for the lack of an atmosphere, Europa is probably similar to Antarctica. Most likely it contains life-generating elements; in the absence of an atmosphere, meteorites can reach the surface without burning up. The only problem is the source of heat. It could come from the Sun, from inside the satellite, or from Jupiter's magnetic field. The heat supplied by these sources is weak compared to what the Earth receives; it could, however, be enough under certain conditions limited in space and time. If I were a consultant at NASA, I would certainly support a project for sending a probe to look for life under the icy crust of Europa.

Titan, Saturn's largest satellite, was thought to be the only satellite in the solar system to have an atmosphere, until Voyager 2 reached Triton. Titan circles the planet in 16 days, at an average distance of 1.2 million kilometers. The two Voyager probes detected nine simple organic molecules in Titan's atmosphere, mainly nitrogen (about 80 percent) and methane (3 to 6 percent), but also propane, acetylene, cyanoacetylene, and carbon dioxide. Abundant protons and other charged particles coming from Saturn's magnetosphere and a thick cloud cover opaque to ultraviolet radiation should favor complex organic chemistry and the production of many different complex organic molecules. Carl Sagan and collaborators irradiated a simulated Titanic atmosphere in the laboratory and obtained a solid organic substance called tholin, which contains about fifteen amino acids of L and D configurations. Further experiments demonstrated that tholin is metabolized as a sole source of carbon and energy by a wide variety of terrestrial microbes, in environments with or without oxygen, thus suggesting that tholin could be a key nutrient source for such organisms on Titan.

If not constantly renewed, the methane in Titan's atmosphere would be entirely decomposed in 10 million years. According to Sagan, an ocean of methane at least 400 meters deep could be its source. However, the temperature profile of the atmosphere obtained by Voyager 1 is incompatible with an ocean of pure methane at mid latitudes, and a global ocean of ethane and methane is ruled out by radar measurements of Titan's surface obtained in 1990.

This ocean would have to be pure methane and limited to high latitudes.

The Cassini mission, a joint project of NASA and the European Space Agency, has received final approval and will be launched toward Saturn in 1996, and should reach it in 2002. ESA will build the probe, which, launched from NASA's orbiter, should slowly descend into Titan's atmosphere to analyze its composition, and once on the ground should rapidly analyze the soil before the radio relay link between the probe and the orbiter fails. In the unanimous opinion of planetologists, Titan's atmosphere should be rich in organic compounds, despite a ground temperature of –179°C! The Voyager probes already detected cyanhydric acid, a forerunner of the purines which are the basis of Earth's amino acids. Titan is probably a vast laboratory on a planetary scale where reactions similar to those which were at the origin of life on Earth have prematurely halted because of very low temperatures.

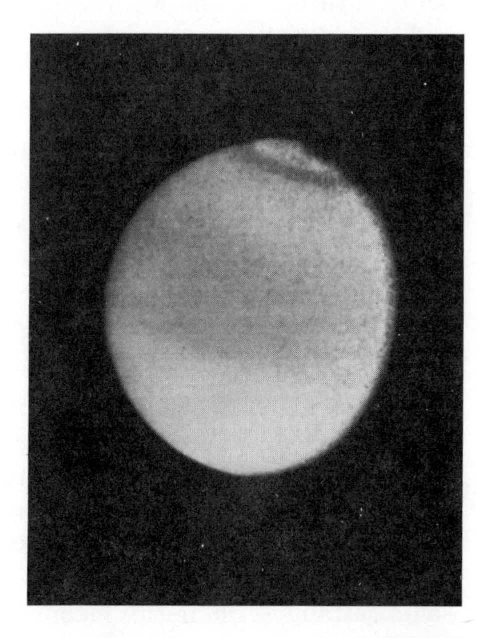

Titan. This satellite of Saturn is the only satellite in the solar system that has an atmosphere. On this picture, taken from 2.3 million kilometers away by the Voyager 2 probe, we can distinguish some details of the cloud system, particularly a dark ring around the north pole.

(NASA)

Iapetus, another of Saturn's satellites, could also be rich in organic compounds produced by the irradiation of methane-rich ice in ultraviolet light. Other bodies in the solar system, such as asteroids, Uranus' rings, and Europa, could also be rich in organic compounds. The origin and the role of the methane and cyanhydric acid (HCN) detected in Pluto's atmosphere should be studied. Methane is also an important component of the atmospheres of Uranus and Neptune. But Titan's atmosphere makes that satellite the most interesting prospect for an extraterrestrial site for the origin of life.

The results of our exploration of the solar system are disappointing. Biological experiments on Mars leave us little hope of finding life on even less hospitable planets. But the extraordinary images transmitted by the Voyager probes — Europa's icy crust, Titan's yellow-orange disk, Triton's volcanoes — reveal once again the inexhaustible variety of nature. The results of forthcoming missions to the satellites of the giant planets and to the asteroids will not fail to surprise us, and will certainly turn up unexpected sites for prebiotic chemistry and perhaps even biology.

We now leave the solar system for what I believe is the real birthplace of the building blocks of life: giant molecular clouds. We already know that stars are born there, and that they contain those molecules which are at the basis of the chemistry of life.

Over the past 15 years, radio astronomers have discovered about sixty molecules in our Galaxy's interstellar medium, and even some in other galaxies. These molecules are concentrated in very massive clouds, obscured by a large quantity of dust and warmer than the surrounding medium. Of the molecules discovered, 65 percent are organic and contain carbon atoms. Let us remember that the human body is formed of chains of organic molecules.

Some of the molecules discovered are very complex; for example, cyanopentacetylene contains thirteen atoms. We can also hope to find more complex molecular structures, such as amino acids or proteins, which are the basis of life's architecture. Such molecules could certainly form in this medium during a gaseous phase under the action of ultraviolet radiation. Although all attempts to find even the simplest amino acid, glycine, have failed, the polymer detected in Halley's Comet by the Giotto probe gives some hope for rapid progress in this domain.

The British astronomers Fred Hoyle and N. C. Wickramasinghe, fervent believers in the panspermia theory, claimed for many years that interstellar dust was formed of bacteria, viruses, and other living cells. This claim was based on analyses of interstellar ultraviolet radiation. Their discoveries were flatly denied in 1984 by a group of researchers at the University of Pennsylvania, led by Robert E. Davies.

It may be surprising that organic molecules can grow in the interstellar environment, which is so hostile to life as we know it. Compared to Earth's atmosphere, it is very thin and cold, because a screen of dust insulates it from stellar radiation. The physical conditions make the organic chemistry in these clouds very different from that on Earth. But dust has an important role to play in the formation of molecules there — a role as important as that of ionized atoms.

This medium is radically transformed during the contraction of the cloud and the star's condensation. Do such fragile and complex molecules survive such changes, and are they the same ones that were present in the Earth's primitive atmosphere? Organic molecules could be collected by comets at the edge of the solar system, where they would have suffered the least during the cloud's contraction. The comets would then bring them into the planets' atmospheres. Interstellar dust could act as a vehicle and as a shield against ultraviolet radiation. I have already mentioned the possible roles played by comets, meteorites, and dust in the specific case of life on Earth.

There is still too much uncertainty about Earth's primitive environment for anyone to be able to show that the first links in the synthesis of life came from molecular clouds and could not have formed on Earth.

Organic molecules are the basis of life, but can they really be considered *signs of life*? We would need a much more spectacular discovery, such as the twenty amino acids linked with life on Earth without any other configurations of amino acids.

We could try to develop a more universal criterion: the presence of abnormal concentrations of molecules. The more complex a molecule is, the more ways there are of assembling its atoms.

In chemistry, the different possible arrangements of a molecule's atoms are called *isomers*. If one day, in a corner of the Galaxy, we were to find a single isomer of a complex organic molecule which has a large number of possible isomers, then we could suspect the interference of a selective process, perhaps one analogous to life. Indeed, the process of converging toward the formation of structures which at first seem completely improbable resembles life very much. To convince ourselves, we would have to go back in time to find the various simpler molecules which contributed to the formation of this isomer. It would be like sailing up a river to look for the streams and smaller rivers that supply it with water.

The search for signs of life could also lead to the analysis of artifacts. The study of the microscopic structure of interstellar dust gathered in space could reveal artificial crystalline forms, unlikely mixtures, or molecular structures that challenge chemistry. Specialists in interplanetary dust have not yet reported finding such structures.

6

The Search for Planetary Systems

If desires embellishes all that it looks upon, the desire of the unknown embellishes the universe.
—Anatole France, *The Book of My Friend*

Life as we know it developed on a modest planet orbiting a star. If there is life somewhere else in the universe, the first target for our research should be planets orbiting other stars. Now, we know of only one planetary system — our own. Of the nine planets in our system, only Earth is known to harbor life.

The search for other planetary systems in our Galaxy can contribute important data to our search for life in the universe. If planets are very rare, the exceptional character of life on Earth will be emphasized. If, on the other hand, all stars are surrounded by planets, asteroids, comets, or disks of dust, we will be encouraged to pursue our quest. Statistical information on the number of planets around stars of different kinds would allow us to better target our search for life in the universe.

Such statistical information is also indispensable for understanding the origin of our solar system and the mechanisms of the formation of stars and planets in general. It would be especially important to know if planets exist around stars of all types, or whether there is a marked preference for solar-type stars. Another important question is: are there planets around multiple-star systems, or only around single stars?

This research is the last stage of the Copernican revolution, in which Earth lost its privileged status as the center of the universe.

If planetary systems are very common, the solar system and Earth become more ordinary. At first, the solar system was considered to be an exceptional phenomenon, owing its existence to a rare event: the passage of a star very close to the Sun tore off a filament of matter, from which the planets formed. However, more recent theories about the solar system's formation assume events which could happen around other stars. If these theories are correct, the Sun is only one of millions of stars in our Galaxy that have planets around them.

The most promising theory appears to be that proposed in 1987 by Takenori Nakano of the University of Kyoto. The starting point in his scenario is the residual gaseous nebula surrounding a star being born. The volume of this nebula encompasses the limits of the future planetary system. The rotation of the protostar has flattened it into a thin disk. If the disk is too massive, it is unstable and rapidly breaks up into gaseous objects much bigger than Jupiter before it can attract solid particles. If, on the contrary, its initial mass is much smaller than the star's, the disk is stable and grows progressively by attracting interstellar grains. Interparticle collisions, very frequent because of turbulent motion inside the nebula, are inelastic because shocks cause the grains to disintegrate rather than bounce off one another.

When the turbulence has lessened, the grains sediment toward the middle of the nebula's disk under the combined effects of their mass, inelastic collisions, and the overall rotation (without which they would fall onto the star). After about a million years they form a very thin and dense disk of dust. The larger grains fall more rapidly, bringing the small ones with them, thus accelerating their fall by a snowball effect. Gravitational forces increase as the disk's mass grows and finally break it up, more or less in the same way that heat makes a plate break. This thin disk then fragments into clumps of dust grains called *planetesimals*. At this stage they are very light — several billion times less massive than the Earth — but their size increases bit by bit under the effect of collisions.

The exact distribution of the disk's mass, whose value we can only guess, determines quite precisely the masses of the various planets and the speed of their formation. The residual gaseous

(La Recherche)

Formation of the solar system. This is a schematic drawing of four steps: clockwise from upper left, the collapse of a protostellar cloud, the formation of a protoplanetary disk, condensation of planetesimals and action of the solar wind, and the current solar system.

nebula still plays an important role by favoring the formation of a very dense atmosphere on the protoplanets. This increases their cross-section, which lets them catch more planetesimals and grow. Radiation and the stellar wind blow this gas out of the stellar system. A very luminous star runs the risk of ionizing[1] the gaseous nebula by heating and of blowing it away before the formation of a disk of dust and its fragmentation. This happens in the case of a star weighing 20 solar masses, around which solid planets cannot form. The maximum mass for the formation of solid planets is 10 times the mass of our Sun.

Protoplanets form progressively by the accumulation and coalescence of planetesimals, starting with those closest to the star. This order of formation explains the relative sizes of different planets: the first ones form from small planetesimals, the next ones from larger planetesimals, and the last ones from large planetesimals in a gas-free environment. In the case of our solar system, Earth would have formed in 2 million years, Jupiter and Saturn in about 50 million years, Uranus in 1.3 billion years, and Neptune in about 4 billion years. Nakano considers that the solar wind would have expelled the gaseous nebula after several dozen million years — that is, between the dates when Jupiter and Saturn were formed. At that point, the planetesimals would have been only 1,000 times lighter than the Earth. The subsequent absence of gas would have considerably slowed the formation of planets more distant from the Sun, and would have reduced their size relative to Jupiter.

In this theory, the rate of formation of a planet located 40 astronomical units from the Sun, the current distance of Pluto, is 7 billion years. We must therefore assume that Pluto was formed somewhere else, for example near Neptune's orbit, and that it was subsequently catapulted into its present orbit. Furthermore, planets could still be forming past Pluto's orbit by coalescence of planetesimals. A hundred astronomical units away from the Sun, these planetesimals would have an average radius of 300 kilometers

1. An ionized atom has lost some of its electrons. It is thus charged and more easily pushed by the electric forces in the stellar wind.

and would be 30,000 times lighter than the Earth. Their mutual collisions could also have formed a disk of small fragments.

Nakano's scenario accounts quite well for planet formation in the solar system, despite a few problems with regard to the most distant planets. If this model is correct, it is a first argument in favor of the existence of other planetary systems in our Galaxy.

Planets are extremely difficult to detect around stars, for technical reasons. In the visible range of the spectrum, a planet is billions of times fainter than a star. The glare of the starlight completely dazzles the instrument which is looking for a fainter object in its neighborhood. In the infrared range, the situation improves a bit; the brightness ratio is "only" 100,000. Searching for planets by radar would be a possibility if the stars closest to us were not 10,000 times further than Pluto. A Soviet astronomer, O. N. Rhziga, has suggested sending radar signals toward these stars in the hope that our technology will have progressed sufficiently to detect echos which may eventually come back to us.

Unable to find planets by direct imaging, astronomers have made an inventory of indirect evidence of their presence. Everything leads us to believe that the solar system is not the only planetary system in our Galaxy.

The two main techniques for indirectly detecting planets are *astrometry* and *spectroscopy*. Astrometry consists in measuring a star's position with respect to a reference point determined by all the neighboring stars in the plane of the sky. By taking two measures of the astrometric position of a star 50 years apart, one obtains its motion in the plane of the sky. Spectroscopy allows us to measure a star's radial velocity (its speed in relation to the Earth) by the Doppler effect. (This is an alteration in a wave's frequency when its source is displaced in relation to the observer. Measuring this alteration gives us the source's radial velocity.) These two techniques allow us to estimate the transverse and radial displacement of stars, through measurement of their position and their radial velocity spread over years.

The point of using these two techniques is to detect small perturbations caused by a small companion in a star's orbit. Indeed, if a star has a planetary companion, the system's center of gravity

Diagram of a rosetta orbit.

is not in the center of the star. The star rotates about this center, and its trajectory in the Galaxy is slightly modified. The amplitude of the perturbations tells us whether we are dealing with a very faint star or a planet. For example, the presence of Jupiter creates variations in the Sun's velocity of about 13 meters per second with a period of 12 years. Such motion can be detected by very precise astrometric measurements of the star's successive positions.

In the solar neighborhood, a dozen stars whose proper motion[2] might be attributed to an obscure companion have been detected. The most famous one is Barnard's Star, which has been studied since 1938 by the Dutch astronomer Peter van de Kamp. He was able to reconstruct its trajectory from 4,000 photographic plates taken over more than 1,000 nights of observation. The oscillations in its trajectory are very weak, several hundredth of an arc second, and could be explained by the presence of two planets the size of Jupiter in circular orbit around the star, with periods of 11.5 and 20 – 25 years respectively. Unfortunately, stability defects in the telescope at the Sproul Observatory, where van de Kamp worked, cast some doubts on the reality of these oscillations. Such measurements are very difficult to make: variations in the

2. A star's proper motion is its motion in the plane of the sky relative to other stars.

Sun's average motion caused by Jupiter would not be detectable 30 light-years away with present techniques. For nearby stars, measurements have to be precise to a 10,000th of an arc second if there is a giant planet to be detected, or to a 10-millionth of an arc second for planets the size of Earth.

The orbital motion of a star also produces variations in its radial velocity with respect to the Earth. Spectroscopic techniques recently perfected at the Canada-France-Hawaii telescope by Bruce Campbell permit detection of variations smaller than 10 meters per second in radial velocity. The stellar accelerometer of Pierre Connes would measure radial velocities to within 1 m/sec. There are a dozen techniques suggested or implemented to measure relative radial velocities to within less than 50 m/sec.

(E. DAVOUST)

The Canada-France-Hawaii telescope. One of the world's ten largest telescopes, it is located at an altitude of 4200 meters, on top of an extinct volcano on the island of Hawaii.

Bruce Campbell has been observing a sample of sixteen stars close to the Sun's type since 1981. Seven of them have velocity variations in the range of 25 – 65 m/sec, which could be explained by the existence of companions slightly more massive than Jupiter. One of these stars, Gamma Cephei, is a double star, but may also have an orbiting planetary companion with a period of revolution of 2.7 years. The six other candidates, which include Epsilon Eridani (10 light-years from the Sun), still have to be observed for a few years before any statements can be made, because the amplitude of the variations in most cases implies a period of revolution of more than 10 years.

There are however many problems in applying these techniques to the detection of planetary systems. First of all, observations must be spread over several years in order to detect variations which presumably have long periods. We do not have a set of standard stars to calibrate the observations; such stars have a radial velocity that is constant and known to within a few m/sec. Finally, even the detection of variations in a star's radial velocity would not necessarily demonstrate the existence of a system of planets. There could be a companion star too dim[3] to be visible with current telescopes, or else there could be variations in the star's atmosphere. Preliminary results should therefore be examined carefully.

The fact that a star spins slowly has long been suggested as an indication of the presence of a planetary system. This indication seems to favor stars like our Sun (G-type stars) or even K or M stars, which rotate much more slowly than other stars. The theory of star formation by collapse of interstellar clouds predicts that these stars originally had a high rotation rate. It is possible that the missing rotation energy has been redistributed to planets. But it is just as possible that another phenomenon, such as a solar wind, dispersed that rotation energy. Slow stellar rotation is thus no longer considered convincing evidence for the existence of planets.

Seventy M-type stars were observed with an extremely precise radial velocity technique by Geoffrey Marcy at Mount Wilson. None of them seemed to have velocity variations compatible with

3. A companion that is too close to be separated on images can still be distinguished by taking spectra, which, in a case of duplicity, show two systems of atomic lines.

the presence of a planetary system, even though they all spin slowly. Furthermore, their mass is much lower than the Sun's, thereby making them more sensitive to the effects of a planetary companion.

Table 7
Rotation velocities of stars as a function of their type.

Type	Percentage	Velocity (km/sec)
O	1	180
B	10	200
A	22	200
F	19	100
G	14	4
K	31	1
M	3	1

Not all stars of our Sun's type are necessarily favorable to the development of life; some form double or multiple systems in tight orbits. In such a system, a planet might undergo very large seasonal temperature variations. Besides, it is not at all certain that multiple stars form with the same string of planets as single stars.

A new and quite unexpected method for finding planets around red giant stars has recently been suggested. Such stars are sometimes surrounded by several sources of radio emission in the millimeter band, due to molecules of silicon monoxide (SiO). These sources are so intense that they can only be masers.[4] Curtis Struck-Marcell suggests that the maser sources are on planets. Their radial velocities, of the order of 10 to 20 km/sec relative to the star, are comparable to those of the planets in our solar system. If this hypothesis is correct, these radial velocities should vary periodically because of the planets' revolution around the star. Furthermore, these masers emit in the millimetric domain, and the new interferometers should be able to measure their motions

4. Masers are molecular clouds which are in a very peculiar excited state, and thus emit very strong radiation, usually at radio wavelengths.

and the new interferometers should be able to measure their motions in the plane of the sky, thus providing the other elements of the planets' orbits.

Other indirect evidence for planetary systems is provided by the fact that "miniature planetary systems" exist around Jupiter, Saturn, and Uranus, each of which has about fifteen satellites. This hierarchical structure in the solar system suggests that stars are not the most elementary building block of the universe.

The first direct evidence for the existence of objects smaller than stars began to appear in the mid-1980s, when, after recent progress in the sensitivity of astronomical receivers in the infrared domain, astronomers reported the discovery of protoplanetary disks and of brown dwarfs[5] around stars.

Astrometric observations made at the US Naval Observatory of the trajectories and distances of a thousand stars close to the Sun allowed astronomers to single out several stars suspected of having planet-sized companions. Followup observations by speckle interferometry[6] in the infrared range were obtained at the Kitt Peak National Observatory in Arizona by Donald W. McCarthy and his collaborators. The star Van Biesbroek 8, located 21 light-years away from us, turned out to have a companion one arc second away — that is, 6 astronomical units. (Jupiter is 5 astronomical units from the Sun.)

Since the mass of Van Biesbroek 8 is only one-tenth that of the Sun, the companion had to be 10 times fainter than the faintest stars in the sky, and almost a million times dimmer than the Sun. Its mass was estimated at 20 to 60 times that of Jupiter, and its effective surface temperature at 1,633°C, which is in keeping with its predicted mass. Such an object could not be a star — the least massive stars must be at least 80 times as massive as Jupiter. Nor could it be a planet like those of the solar system. It had to be something in between: a failed star. Research on brown dwarfs was now beginning.

5. The term "brown dwarf" was coined by the American astronomer Jill Tarter.

6. This technique allows one to get rid of the atmosphere's blurring effect by taking numerous instantaneous images of the object and then recombining them after a particular image processing.

Van Biesbroek 8 was observed again in 1986 by Christian Perrier and Jean-Marc Mariotti, both of Lyons Observatory, using the same interferometric method. Even looking as close as a quarter of an arc second from the star, they found no companions. This cast doubt on the observations by McCarthy and his colleagues. The latter redid their observations and did not find the companion either.

This kind of misadventure — scientists finding too hastily what they were hoping to find — is quite common in fundamental research. This one had at least some positive result: that of initiating theoretical research on the nature of hypothetical brown dwarfs, and stimulating observational programs to find similar objects. The estimates of temperature, size, and mass derived from the presumed luminosity of Van Biesbroek 8's companion remain valid for other possible brown dwarfs.

What actually is a brown dwarf? It is an object at the uncertain frontier between star and planet, with a mass 6 to 10 percent that of the Sun and a temperature of about 2,000 to 3,000°C. It is a star that failed to ignite for lack of mass. Its high surface temperature does not come from internal nuclear burning, as in a classical star, but from the energy produced by its slow contraction. Jupiter and Saturn also have an internal energy of identical origin; however, it raises their temperature only about 80°C.

Without getting discouraged, McCarthy continued his observations. In August 1987 he announced finding a brown dwarf around the red dwarf star Gliese 623, 24 light-years from the Sun. This dwarf's mass would be 80 times that of Jupiter. The red dwarf was known to have a companion ever since 1978, but 5 years' observations were needed to determine its orbit in order to calculate its mass by Kepler's laws.

Another group of American infrared astronomers think they too discovered a brown dwarf, this time around a white dwarf star, GC 29-38. They did not use interferometry. They measured the star's total flux at six wavelengths in the infrared range, and compared their observations with the theoretical flux expected for a star of this type. This discovery, published in 1987, followed two fruitless attempts to discover brown dwarfs around fourteen white dwarf stars. The point of observing white dwarf stars is technical: they emit little light in the infrared range, thus making detection of a possible dim companion easier. The companion of star GC 29-38[7] would be a brown dwarf whose surface temperature is estimated at 1,500°C. This corresponds to a radius equal to 15 percent of the Sun's and a mass between 4 and 8 percent of the Sun's.

Among the few brown dwarfs documented in the literature, the companion of the white dwarf GD 165 detected at a wavelength of 2 microns is the most credible. The authors of this discovery believe that there could be many brown dwarfs in our Galaxy. These are, however, very difficult to find, because a brown dwarf as close as Neptune would appear no brighter than Mars.

What should we think of the star HD 114,762, which has recently made the headlines in scientific magazines? The radial velocity measurements, some of which were obtained at Swiss telescope of the Observatoire de Haute-Provence with the Coravel instrument, revealed variations of 726 m/sec with a period of 84 days in this solar-type star located 90 light-years away from us. Such variations could be caused by a planet at least 10 times more massive than Jupiter in a very tight orbit around the star. It might be a brown dwarf rather than a planet, but this takes nothing away from the importance of this discovery.

Moreover, Swiss astronomer Michel Mayor and his team detected other brown dwarfs by this kinematic method. In a sample of 166 G-type stars close to the Sun, they found that only 18 percent were probably genuine single stars. The majority (65 percent) have a companion with at least one-tenth as much mass as the star, and about 18 percent could have a planetary companion. They also offer a criterion for telling brown dwarfs from planets: the former have eccentric orbits, while the latter have almost circular orbits. Photometric and kinematic methods for detecting substellar objects should obviously be combined. We would then be able to draw a fairly precise picture of the astronomy of these objects in the near future.

7. also known as WD 2326+049, ZZ Psc, Gl 895.2, EG 159, and LTT 16907, to confuse astronomers!

Brown dwarfs are not really planets, because they are tens of times more massive than the biggest planet in the solar system. But, if confirmed, their discovery is a first step toward the search for dim objects around stars.

Another step in this direction is the discovery since 1983 of many cases of excessive infrared radiation around stars. In seven cases, this excess of radiation is most likely due to protoplanetary disks, in other words nascent planetary systems.

In 1983, scientists at the University of Hawaii discovered an infrared source around the very young star[8] T Tauri by speckle interferometry. Since then, a study of a sample of 59 young, low-mass stars of the same type in the stellar complex Taurus-Auriga revealed that almost 20 percent of them emitted excessive infrared radiation. The idea that this emission could come from a giant condensing protoplanet has now been abandoned. Its source is more likely to be protoplanetary disks.

In the same year, Japanese radio astronomers discovered what they thought were huge protoplanetary disks — several tens of thousands of astronomical units in diameter — around a star in Orion and the star IRS 5. They observed emissions of the carbon monosulfate molecule (CS) at millimeter wavelength. Another very young star, HL Tauri, 500 light-years from the Sun, is probably also surrounded by a giant molecular disk several thousand astronomical units in diameter. Images obtained by speckle interferometry in the infrared range reveal an elongated halo, but also ejection of matter. More recent observations of two of these stars (see table 8) showed that these dust rings are much smaller than the disks observed by radio. The latter might be the remainder of the initial protoplanetary nebula.

The astronomical satellite IRAS was launched into orbit in 1983 to measure infrared cosmic radiations. It revealed that 20 percent of the stars in the Main Sequence[9] have an infrared excess, probably caused by protoplanetary material. If we assume that this excess corresponds to an evolutionary phase through which all

8. On an astronomical scale, that is several hundred million years.

9. The Main Sequence is the evolutionary stage of a star during which the nuclear fusion of hydrogen into helium takes place. The Sun is midway through this phase.

stars in the Main Sequence pass, we can estimate that this phase lasts 20 percent of the average star's life — about a hundred million years. This agrees with estimates of the time needed to form planets.

From this point of view, the Sun would have no significant infrared excess if it was seen from far enough away for an observer to observe the radiation from the whole solar system at once. This means that many planetary systems could be missed for lack of resolution.

IRAS detected large infrared sources around the stars Vega and Fomalhaut, both of which are about 20 light-years away from the Sun. These sources could be a ring of dust 80 to 100 astronomical units in diameter surrounding each star. Vega, a very luminous star, evolves 30 times more rapidly than the Sun. This is probably too fast for a planetary system to form and for an elaborate form of life to develop. Fomalhaut is less luminous and should live longer. The observations of Vega were confirmed in 1984 by American astronomers observing from an airplane at very high altitude; they found, however, that the disk was twice as wide. It is difficult to estimate the total mass of dust around the two stars, but perhaps there is enough to produce at least a small planet. We should also keep in mind that the giant planets that make up 90 percent of the solar system's mass are mainly composed of hydrogen and light elements, which are invisible in the infrared range.

Table 8
Protoplanetary disks discovered since 1983.

	Age (millions of years)	Distance (light years)	Radius (astronomical units)	Mass (Earth masses)	Observatory
Vega	500	25	85	300	IRAS, airplane
Fomalhaut	500	22	50		IRAS
HL Tauri	1	450	160	1	Kitt Peak and
R Monocerotis	1	1850	500	5	Hawaii
Epsilon Eridani		9			IRAS
Beta Pictoris	500	50	214		IRAS, Chile
IRS 5	1	450	250	18	Hawaii

From table 8 we can tentatively state that there are two kinds of disks: residuals from the formation of planets around young stars, and protoplanetary disks around very young stars (such as HL Tauri and Monocerotis). The latter seem much lighter because they are essentially made of gas.

Observation of these disks does not prove the existence of planets. The disk could very well disperse into the interstellar medium, perhaps by being pushed by the stellar wind coming from the star, instead of condensing into planets. The causal relation between disk and planet still has to be studied. But these observations bring us into a new era of astronomical research, that of extrasolar planetary systems.

Astronomers realize that planets are not detectable in the infrared range with current telescopes because of their small surface-to-volume ratio. Furthermore, the Sun's zodiacal dust would frustrate any effort to see Jupiter with current telescopes from a neighboring star. This is why so much hope is placed in space-based telescopes.

The Hubble Space Telescope, now in orbit 500 kilometers above the Earth, will contribute to planetary research, since four of the five instruments on board may be used for this purpose. The exceptional resolution of its imaging instruments will be capable of resolving a sun-planet system up to 33 light-years away from the Sun. One possible procedure is the following: measure positions of 100 nearby stars for 5 to 10 years with the fine guidance sensors, to detect epicyclic orbits; measure radial velocities of the detected planets with the high-resolution spectrograph to determine their motion, their atmosphere, or an infrared excess; detect the transit of the planet in front of the star's disk by fast photometry at the time of conjunction; take pictures of the planet when it is farthest from the star with the Faint Object Camera and a coronograph.[10]

The scientists who participated in defining and building the instruments on board the space telescope will have priority in future observations. The 307 programs for which they submitted observing proposals will be protected with provisions like patents

10. A coronograph is a device to stop starlight from reaching the receiver.

(NASA)

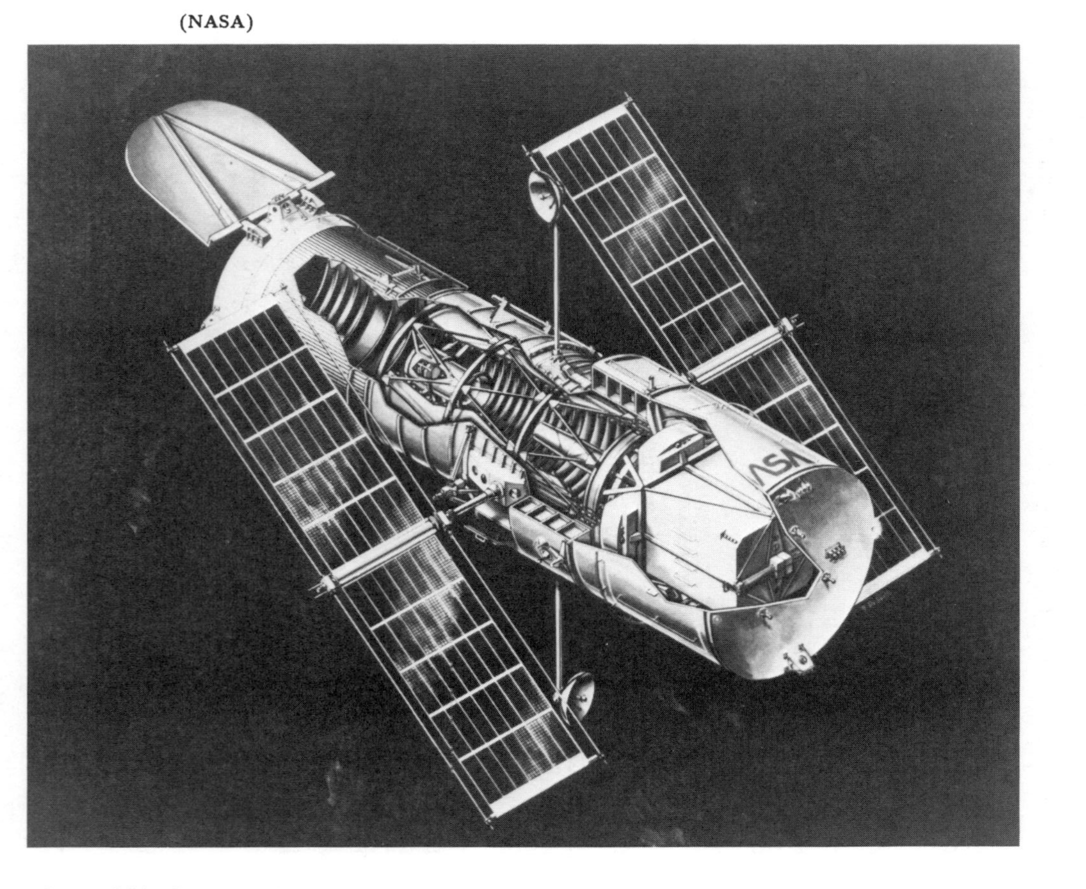

The Hubble Space Telescope. The cylinder containing the telescope and the scientific instruments is 14 meters long and 5 meters in diameter. A small radio antenna receives orders for piloting the instruments. Solar panels collect for 30 months. Among these programs, six concern the search for planets around stars and three concern circumstellar material (such as protoplanetary disks).

The remarkable qualities of this telescope, especially in the area of angular resolution, forecast great progress in planetary research when the second-generation instruments are installed in 1993 or 1994. Until then, high-resolution-imaging observations will concern only bright celestial objects, because of the defects (spherical aberration) detected in the optical system after launch.

The Astrometric Telescope Facility (ATF), a cooperative project between NASA and the University of Arizona, will be placed aboard the American space station to be assembled in space between 1995 and 1999. This instrument will first be used to search for relatively heavy planets the size of Uranus or Neptune. Later it will be capable of discovering planets like Earth 30 light-years away from us. But in order for the results to be reliable, observations will have to be made over 12 to 15 years.

Another NASA project, the Infrared Telescope Facility (ITF), will have a mirror 80 centimeters in diameter and will be able to detect a planet like Jupiter as far away as Barnard's Star in 3 seconds.

All these research projects leave us expectant, although considerable progress has already been made in the past few years. If I had written this chapter before 1983, it would have been pretty short. The recent discoveries of brown dwarfs and protoplanetary disks, and the extremely precise measurements of the motion of nearby stars, show that we are arriving at the threshold of a new era in astronomy — that of substellar objects. Planetary systems are surely not rare in our Galaxy.

But discovering planets does not mean that we have found life. Even if we soon discover planets around other stars, which I feel is likely, the detection of signs of life may be out of reach for another decade or more. The search for life outside the solar system remains an apparently utopian undertaking.

Having left the solar system, we have to abandon the restrictive hypothesis of Earth-type life and consider as a living entity any system that has reached a certain level of complexity. The existence of such systems is probably linked to a physicochemical imbalance in their immediate environment. We then have to find, for example, signs of disequilibrium in planetary atmospheres, such as the excess of oxygen in the Earth's atmosphere. But when can we hope to measure such a disequilibrium in the atmospheres of planets which we technically still cannot detect? As was suggested in the preceding chapter, we should look for signs that could be forerunners of the birth of life — such as an abundance of chains of complex molecules and the proportion of different isomers[11] in protoplanetary or protostellar nebulas.

Another task that is already manageable is to find out whether there are continuously habitable zones in planetary system — zones where the environment remains stable long enough for complexity to appear, and where basic energetic requirements are met.

11. Isomers are different geometrical arrangements of atoms to form the same molecule.

Intelligent Life in the Universe

7

Demography of Extraterrestrial Civilizations

There is no doubt that there is an invisible world. However, we can wonder how far it is from downtown and how late it stays open.
—Woody Allen, *God, Shakespeare and I*

The question of life in the universe becomes really interesting only when it leads to another question — that of *intelligent* life in the universe. Indeed, if it is useful to know that we are perhaps only one form of life, it is much more important for us to know whether we are the only intelligent beings in the universe. The philosophical, social, and maybe even technological implications are much deeper. A possible dialogue with other intelligent civilizations in our Galaxy would have repercussions in every aspect of our existence. On the other hand, the absence of any such civilization would force us to examine the purpose of our own existence in a universe where we are unique.

In the following chapters, I will discuss the possible existence of extraterrestrial civilizations. But what do I mean by "civilization"? How much can the terrestrial model help us? Soviet astronomers, responding to a survey in 1972, offered some definitions that provide a good starting point:

- A very stable state of matter, capable of gathering information, analyzing by abstraction, and using the greatest amount of information about its environment in order to know it better and protect itself. A civilization is defined by the volume of information accumulated, by the way it plans to function, and by its carrying out those plans.

- A very complex machine, capable of keeping and transforming information, able to analyze a situation and using the results of this analysis to direct its evolution in the framework of oriented activities.
- Any system capable of surviving in its environment, and able to keep and transform information in coded form in order to accomplish logical operations and to use this knowledge in directed activity.
- A highly organized system, active on at least a planetary scale, capable of having a directed activity, knowing the surrounding world and itself; in other words, able to make a model of the world using abstract concepts, using its knowledge to reach its purpose.
- A group of interacting subjects asking questions about the existence of other gifted groups.
- Any state of matter capable of emitting sensible information into its environment.
- A system of procedures and mechanisms allowing an adaptative activity which is not genetically programmed. The essence of a civilization, i.e. living beings united in a stable collectivity, is the ability to produce a non-biological organization of this collectivity and the means of adapting to one's environment. This organization allows the activity of the individuals.
- Groups of intelligent beings using the exchange of information, energy, and matter to accomplish their activities, having specific means of monitoring life and ensuring its progressive development.

These definitions reveal more what different societies on Earth have in common — the need to manage information about the environment in order to survive, and to communicate — than the properties of hypothetical extraterrestrial civilizations. This anthropomorphic point of view is inevitable because we don't know any societies other than those on our planet. Such a point of view is not entirely unjustified, since at least minimal resemblance is necessary to make any possible contact interesting.

We can exercise our imagination in generalizing this concept of civilization. First of all, the chemical reactions which govern life on Earth and specifically our perceptions of time and space set meaningful time scales for us (between one second and several decades), and distance scales (between one millimeter and a few hundred kilometers). By civilization, I mean a group of individuals who have a specific purpose and who interact. These individuals could be more or less autonomous. We could thus imagine a society composed of individuals similar to the atoms which make up the human body — a collective organism.

Earth itself could be considered a living organism, or a *geophysiological* system. In 1972 the geophysicist James Lovelock called this the Gaia theory, from the name of the goddess of the Earth in Greek mythology. He believes that evolution of living species, the atmosphere, and the Earth's crust are inseparable. The idea that living organisms can have a regulating effect on climate and on the atmosphere's composition is contrary to the orthodox theory of evolution, which claims that species only adapt to given conditions and have no foreknowledge of their effect on climate. However, a self-regulating planet would explain the relative stability of our environment for the past several thousand years.

The Gaia theory — the theory of the living planet — is mainly a lovely metaphor for the time being. It has, however, been gaining ground among scientists in different fields who are now beginning to search for the global causes of the remarkable equilibrium which allows us to live. We will return to this point in a moment.

Now let's imagine life based on nuclear reactions. Elementary particles the size of protons or electrons could form groups sufficiently complex to exhibit intelligent behavior. In this case, the scales of time and distance would be considerably reduced, since nuclear reactions cover very short distances and thus happen in extremely short intervals of time. Such a civilization could last less than a minute.

At the other end of the cosmic scale, intelligent life could be ruled by complex gravitational interactions and appear as giant gas clouds containing a hundred or a thousand times more matter than the Earth. In this case, humans would play the role of atoms. Beyond that, imagine that galaxies themselves, with their hundreds of billions of stars, have reached a degree of complexity like that of life. Life spans would then be counted in millions of years, and distances in thousands of light-years.

How would information be managed and stored in these hypothetical cases? On Earth, this is done by DNA and the brain. But these speculations are not gratuitous. In fact they show that a reasonable degree of anthropomorphism is justified when we invent search strategies and when we put ourselves in the place of possible extraterrestrial civilizations in order to imagine their behavior. We could establish interesting contact with an extraterrestrial civilization only if it resembled us at least a little bit. After all, what do we have in common with a galaxy or an elementary particle? What hope would we have of someday communicating with such life forms? Without precisely defining the degree of resemblance, I think that having life spans and dimensions of the same order of magnitude are minimal requirements.

The next step in the search for life — deciding whether a planet is continuously habitable — has been made possible by recent progress in the fields of climatology and geosphere-biosphere interaction on Earth. We have to find the zone within which the atmosphere of a planet like Earth would remain stable for several billion years.

The answer to the question of habitability strictly depends on the definition of life. The range of temperatures and pressures in which living organisms are encountered is very wide, but the most complex ones will survive only in a narrow interval of temperatures. This more restricted interval should be considered as most favorable for the survival of life and its evolution toward more complex forms.

Why did life develop on Earth and not on the neighboring planets? Will industrial and domestic pollution of the Earth's atmosphere soon create a greenhouse effect like that on Venus, where the temperature on the ground is 500° C? Does life play a regulating role in the atmosphere's composition and temperature, as the Gaia theory proposes? Would a nuclear winter necessarily lead to the extinction of life? These questions are beginning to find answers, thanks to computers which allow us to deal with global climatic models within a reasonable length of time.

A relatively stable environment and a climate variable over limited zones are indispensable to the maintenance of life. In 1971, Michael Hart, then a researcher at NASA, estimated the continuously habitable zones for earthly life.[1] He found a very narrow band around the Sun. In fact, if the distance from Earth to the Sun changed by only a small percentage, the climatic and ecological consequences would be catastrophic.

If the Earth came closer to the Sun, it would heat up. As water evaporated from seas and other bodies of water, it would produce an atmosphere rich in water vapor, able to efficiently absorb the Sun's infrared radiation. Through this greenhouse effect, the atmosphere's temperature would increase several hundred degrees. This would have happened had the Earth formed at a distance 5 percent closer to the Sun 4 billion years ago. If, on the other, the Earth-Sun distance had been 1 percent greater 2 billion years ago, when the atmosphere oxygenized, the Earth would have been a frozen planet, like Jupiter's satellites. Glaciation is difficult to reverse; it would remain until the distance from the Earth to the Sun decreased at least 10 to 20 percent (impossible because of the dynamical stability of the solar system), or until the Sun became a red giant.

The distance limits above are valid for the Earth and the Sun. For a star which is less massive and therefore less luminous than the Sun, the continuously habitable zone is closer to the star, but it is also narrower. According to Michael Hart, the zone disappears completely if the star's mass is more than 17 percent smaller than the Sun's. This eliminates all M-type stars and most K stars.

All these results convinced Hart that life is probably not very widespread in the universe, and he has since become one of the advocates of a silent universe. But we must also take into account a planet's geology and biology when estimating continuously habitable zones. Climatic conditions depend on the atmosphere's composition, which in turn depends on geology and (to a lesser extent) on the living world.

---

1. Continuously habitable zones are the regions in a planetary system that can be permanently inhabited by living organisms. They are narrower than zones habitable at a given period because of slow variations in the star's brightness and the evolution of planetary atmospheres.

(Scientific American)

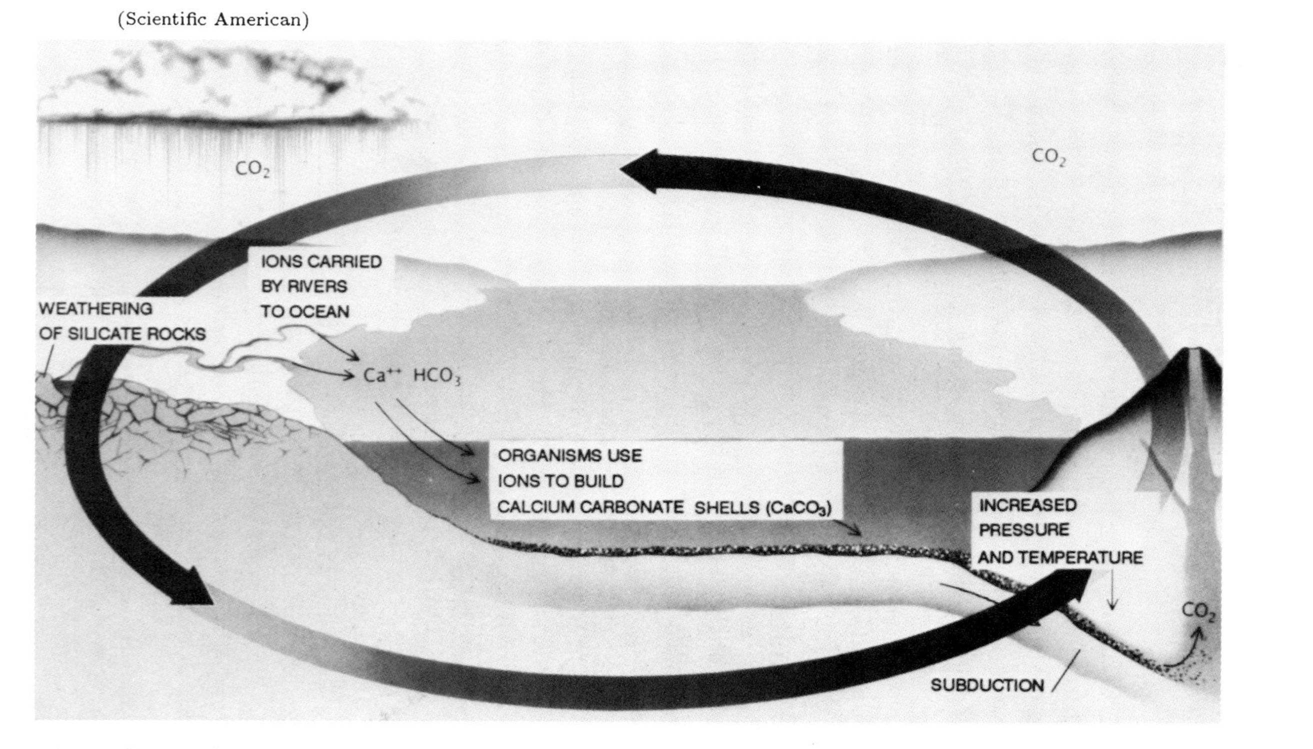

The carbonate/silicate cycle. Carbon dioxide probably plays an important role in regulating the temperature of Earth's atmosphere, as this diagram shows. Global temperature rises with an increase in the rate of carbon dioxide in the atmosphere, but rains resulting from the evaporation of the oceans wash the gas out of the air.

In 1988, three other researchers at NASA proposed a climatic model in which the continuously habitable zone of our solar system extends from Earth's orbit to that of Mars. To prevent any runaway glaciation or greenhouse effect, there must be a temperature-regulating mechanism. James Pollack and his colleagues suggested that the exchange of carbon dioxide between the atmosphere and the Earth's crust plays this role, with a cycle which can speed up or slow down depending on the temperature. When it is too hot, ocean water evaporates and more frequent rains wash some of the carbon dioxide out of the air, thereby reducing the greenhouse effect. The carbon dioxide returns to the oceans in the form of carbonic acid, which plankton transform into calcium carbonate. This carbonate sediments and finally becomes incorporated into the Earth's crust. Under the effects of heat and high pressure, it reacts with silicates to produce new carbon dioxide, which escapes gradually (or abruptly, during volcanic eruptions). If, on the other hand, it is too cold and the oceans freeze, there is no more evaporation and nothing stops carbon dioxide from accumulating in the atmosphere. Under these conditions, the atmospheric temperature can increase by 50°C in 20 million years — enough to make the ice melt.

But why didn't this same process work on Venus and Mars? If those planets formed from the same planetesimals as Earth, their geological composition should be the same. The problem with Mars seems to be its mass, which is only one-tenth that of the Earth. Mars' high surface-to-volume ratio made it cool too quickly, thereby preventing the carbon dioxide from escaping from carbon rocks. Venus' problem is more serious. The layer of cold air that on Earth stops clouds from rising into the stratosphere does not exist on Venus. The oceans thus evaporated irreversibly in several million years, eliminating one of the regulating factors.

The results of the "general circulation models" must nonetheless be considered with care because of the many uncertain or unknown parameters. Present theories do not allow us to predict whether variations in cloud coverage on a planet's surface have a positive, a negative, or a neutral effect on the global temperature. The parameters of heat transfer between the equator and the poles are also uncertain. Continental drift, variations in the extent of the continental surface (which has been 20 percent on Earth in the past 180 million years), and ocean currents all have some effect on the climate, and this complicates the task of theoreticians. In climatic models, it is impossible to take into account the *a priori* unknown continental configuration of a planet or its possible modifications. The biological element, absent from these models, is also important, at least on our planet.

Geosphere-Biosphere, an international program, was launched in 1986 by the International Council of Scientific Unions to study the dynamics of the whole Earth, taking all its components into account. The main motivation for the program is the possible threat of a greenhouse effect, whose extent and multiple consequences, especially for life and human activity, need to be evaluated. Ocean-atmosphere interaction, the influence of oceanic circulation on the climate, and the role of forests and plankton in regulating the proportion of carbon dioxide in the atmosphere and its temperature are among the principal themes now being studied.

Among the biological agents which could have a regulating effect on the global climate, the forests are primary. Through photosynthesis of chlorophyll, they produce most of the oxygen we breathe. Forest fires seem to adjust the oxygen rate in the atmosphere in a cycle linking the ashes to phosphorus and oxygen dissolved in oceans. They are also indispensable for maintaining vigor in forests, getting rid of underbrush and dead wood and regenerating soil with the ashes. Isn't it amazing that many kinds of trees spend up to half their energy producing isoprene, a highly flammable substance whose only role seems to be encouraging forest fires?

Ocean phytoplankton could also intervene in the balance between biological, climatic, and geological processes on our planet. They produce a gas which provokes condensation of clouds and thus lowers temperature. But the plankton do not seem to produce this gas in direct response to a global temperature increase, and we do not yet know how they could moderate a temperature decrease.

This cooperation of the various components of the environment in order to maintain it, surprising though it may seem at first, could be the result of a process of apprenticeship and mutual adaptation. A self-regulating planet would at least explain the relative stability of our environment over millions of years, making up for the effects of a progressive increase of 25 percent in the brightness of the Sun since its formation. It would, especially, explain life's role in the formation of the environment. (If there is regulation, it is not perfect; there have always been important oscillations between ice ages and milder periods.)

This more global approach to the physicochemical activity on the surface of our planet could be profitably applied to other existing or hypothetical planets.

To the zones made continuously habitable by solar or stellar heating, we must add zones around giant planets or their satellites which are heated through tidal effects. Under its icy crust, Europa certainly has oceans. The heat resulting from the friction caused by tidal forces within the satellite would prevent the water from solidifying. How can we otherwise explain the absence of craters on Europa's surface? As was mentioned earlier in this book, these oceans could be a favorable medium for the development of a form of life comparable to that on the primitive Earth.

Conversely, the presence of a fairly massive satellite could favor the development of life on a planet. Earth's tides, essentially caused by the Moon, allowed life to leave the oceans. The Moon may also have played a role in building the magnetic field that surrounds the Earth and protects us from cosmic rays. But in "double planets" (two planets, or a planet with a large satellite) other determining factors could intervene against life. For example, the Pluto-Charon system is certainly too light to retain a dense atmosphere and too far from the Sun to receive significant amounts of solar energy.

If a planet's position relative to its star plays an important role in the appearance and development of life, the star's life span is also important. A star like the Sun remains stable for about 10 billion years, during which time hydrogen changes into helium in the center of the star. Next, helium nuclear burning, when the star becomes a red giant, is preceded by cataclysms which can endanger all forms of life unless this life can escape far enough away from the star. A more massive star evolves more rapidly. If it is 15 percent more massive than the Sun, its life span is only 5 billion years — half as long — before it becomes a red giant. In such a star the continuously habitable zone would be wider, but it would not last as long and thus life might not reach a very high level of complexity.

On the other hand, a star less massive than the Sun would live much longer before going on to the cataclysmic stage. But only a planet very close to the star would receive enough energy for an Earth-type life to develop there. On a planet too close to the star, the tidal forces induced by the star would have devastating effects. The planet might find itself locked in a synchronous orbit around the star, like the Moon around Earth, always turning the same side to the source of light and energy. There would be no day-night alternation, and no seasons. This, then, eliminates stars whose mass is less than 70 percent that of the Sun.

It seems, therefore, that only stars whose mass is close (within 20 percent) to that of the Sun should be considered as being favorable for the development of life on a planet. But the dimensions of habitable zones around these stars are difficult to estimate precisely, since we have no exact information on the geology of possible extrasolar planets.

Can we assert that multiple stars do not have habitable zones? Computer calculations of orbits show that there are stable circular trajectories around binary stars. But if the separation between the stars is less than 3 astronomical units, the collision speed between rocky fragments is too high for planets to form. As for systems of three stars or more, their complex orbits could critically reduce possible habitable zones. In view of the uncertainties, astronomers prefer to eliminate multiple stars from their statistics.

Finally, "habitable zone" does not mean "inhabited zone." Even if all the necessary conditions for life to appear are met, it takes complete confidence in the deterministic laws of physics and chemistry to assert that life will appear sooner or later. Although all the above-mentioned chemical processes on Earth have had the time to occur innumerable times across the universe over the 15 billion years of its existence, the specific class of biological processes is so exceptional in the eyes of biologists that they categorically reject the idea of life widespread throughout the universe. Despite these restrictive hypotheses, there is still much uncertainty in enumerating the possible sites for life.

I would now like to try to evaluate the probable number of intelligent civilizations that resemble ours and to situate them in the Galaxy. My attempt at demography is based on the assumptions that the life forms that interest us need a source of constant stellar energy for billions of years to reach a stage of evolution similar to our own and that they develop on planets in orbit around stars. Such a census is therefore limited to extraterrestrial civilizations occurring in situations a lot like ours.

To estimate the probable number of civilizations in our galaxy, I will resort to Drake's equation, named for the first astronomer who tried to detect artificial radio signals of extraterrestrial origin. It is also called the Green Bank equation, for the American radio astronomy observatory where this first attempt was made. This equation — simply a product of probabilities — has often been used by astronomers searching for extraterrestrial life:

$$N = RF_* N_p F_h F_l F_c L$$

where

N = number of intelligent civilizations in Galaxy
R = mean rate of star formation in Galaxy
F_* = fraction of stars with habitable ecosphere
N_p = number of planets around these stars
F_h = fraction of habitable planets
F_l = fraction of habitable planets with life
F_c = fraction of planets with intelligent civilizations
L = longevity of a civilization.

R, the mean rate of star formation in the Galaxy, is of the order of 10 to 20 stars per year, if we divide the number of stars in the Galaxy (200 billion) by the age of the universe (between 10 and 20 billion years). F_*, the fraction of stars with habitable ecospheres, is 0.14 if we count only G-type stars (see table 7), but the number has to be reduced by a factor of 5 if we count only G stars likely to have planets, following recent studies by Swiss astronomers mentioned in the preceding chapter. N_p, the number of planets around these stars, is 9 in the only known case, the solar system; in the same case F_h, the fraction of habitable planets, is 0.11 — only one planet out of 9. F_l, the fraction of inhabited planets, and F_c, the fraction of these planets with intelligent civilizations, are unknown but are less than 1 by definition. As for L, the longevity of a civilization, according to the wisdom of civilizations and their resistance to internal crises, this factor can vary from 100 years to 1 million or more.

The product of all these factors gives an estimate of N, the number of intelligent civilizations in our Galaxy. Since these parameters are mostly unknown, the value of N differs greatly according to whether we are optimistic or pessimistic. In the worst case, $N=1$, we are alone. In the best case, N can reach several millions. Some astronomers take into account temporal variations in the parameters of Drake's equation when estimating N. The rate of star formation in the Galaxy was indeed higher in the past, and technological civilizations can have varying life spans, just like human beings. But these refinements can in no way reduce the uncertainty over N; only concrete results in detecting intelligent civilizations can do that.

Drake's equation does not furnish any valid quantitative estimate of the probable number of extraterrestrial civilizations in the Galaxy. Rather, it shows us the difficulties and unknowns that still need to be cleared up before serious demographic work can be undertaken.

However, there may be cause for optimism. If the life spans of technological civilizations are very short, N is very probably less than the total number of civilizations which have developed and disappeared since the beginning of the universe. But the probability of detecting signs of past intelligent life might remain high if these civilizations left permanent visible artifacts or audible (radio beacons) traces of their existence.

A second reason for being optimistic lies in the hypothesis that extraterrestrial civilizations sooner or later undertake to colonize their environment. The Galaxy may be sprinkled with small galactic empires which developed around parent stars. In this case, our search at the telescope should not limit itself to targets similar to the Sun. According to calculations by a group of chemists at the University of Maryland, our chances of discovering "them" are enhanced by a factor of 10 if galactic colonization has started, all else being equal.

Table 9
Mean distance in light-years between civilizations as a function of their number in the Galaxy. The third row gives the probable number of stars we must observe to have a chance of discovering an emitting star.

Number of civilizations	Distance between civilizations	Number of stars for one civilization
10	20,400	20,000,000,000
1,000	4,400	200,000,000
100,000	1,000	2,000,000
1,000,000	440	200,000
200,000,000	44	200

Table 9 is intended to illustrate in another way the difficulties astronomers face in their search for artificial signals of extraterrestrial origin. For example, the very optimistic estimate of one million intelligent civilizations in the Galaxy represents only one inhabited star out of 200,000. If we have no preconceived ideas about the probability of a star's being hospitable, we have to "listen" to at least 200,000 stars to be sure to have listened to one that is inhabited. And this star would be at least 440 light-years from us. It would take 880 years from the time a message is emitted for a response to be received.

The above estimates have statistical value — they are averages on a galactic scale. But galactic population density is certainly not uniform if it is not nil. Just as humans gather in sites favorable to life and avoid deserts, it is very probable that extraterrestrial civilizations prefer to be in certain areas of the Galaxy.

Stars are more numerous in the galactic disk, and especially in the spiral arms. The Sun is located on the edge of the Orion arm.

Globular clusters and the galactic center are other areas of heavy stellar density. But the Galaxy's spiral arms are mostly formed of stars which are too young to have advanced life. On the other hand, the globular clusters are formed of stars that are too old to contain metals, which are necessary for the structure of living beings. As for the center of the Galaxy, it is the location of violent events.

If the most advanced extraterrestrial civilizations have means of emigrating to sites more favorable to their development, we should probably look for them near galactic sources of natural energy and far from areas of catastrophe. The Galaxy is full of dangers for systems which need a stable environment: supernova explosions, black holes, variable stars, magnetic stars, some pairs of stars in tight orbits, sources of x rays and gamma rays, and so on.

Stars are the most obvious sources of energy, but pulsars could also be interesting ones. They are rotating neutron stars which periodically emit intense bursts of radio waves, the way a lighthouse emits light beams. Pulsars could also be useful for interstellar communication if their pulsations could be modified artificially.

When Earth formed, the universe had already existed for at least 5 billion years. This may suggest that life on Earth must be young relative to the life around much older stars. But the heavy elements on which life is based (carbon, oxygen, nitrogen, etc.) did not exist at the origin of the universe; they formed progressively in stars. So we must search for life around second-generation stars, formed with their planets from gases rejected by the first stars.

There are other selective criteria in the search for extraterrestrial civilizations. I will come back to them in the discussion about searching for artificial radio signals.

8

Sociology of Extraterrestrial Civilizations

But if our vision stops there, our imagination should go beyond.
—Blaise Pascal, *Pensées*

"But where are they?" The nuclear physicist Enrico Fermi asked this question in 1946. The question reveals the paradox between our painstaking research and the absence of extraterrestrial signs. If there really are extraterrestrial civilizations, why have they not yet demonstrated their existence to us?

At the time when the question was raised, almost everything that had been written on the subject of extraterrestrial civilizations was science fiction. Numerous scientific works have since been written to estimate the probable number of extraterrestrial civilizations in our galaxy, or to describe the most efficient means of detecting them and of communicating with them. The many attempts at detecting radio signals of artificial origin have remained fruitless, and the questions asked about this great cosmic silence have become more and more pressing. Astronomers have had to resort to risky exercises in extraterrestrial sociology to explain it.

The paradox vanishes if we accept the idea that we are in fact the only intelligent life form in the universe. But scientific progress since the sixteenth century leads us to the opposite belief. We have successively learned that the Earth is not at the center of the universe, or even at the center of the solar system, and that the Sun is an ordinary star among hundreds of billions of other stars in our galaxy, which in turn is only one among billions of galaxies.

Each step in the development of our knowledge has been a lesson in modesty. The logical continuation of this progress would be to discover that there are millions of planets like the Earth in our galaxy, and that we are only one of innumerable life forms that crowd it. This point of view, as well as the importance of the question, explains astronomers' perseverance.

This perseverance is based on elements which must be taken seriously since they are essential to the very definition of a human being. Intellectual curiosity, for example, is one of the principal qualities of an intelligent being. Another is communication — a fundamental need of humanity. We don't need to extrapolate much to think that, if there are other civilizations in our Galaxy, they would be motivated to explore their environment. This is what man has done, exploring the Earth, searching the sky with telescopes, then sending exploratory probes to the Moon and to Mars and the other planets in the solar system. Only the relative slowness of our probes (Voyager needed 4 years to reach Saturn, 12 years to reach Neptune) prevents us from going further for the time being.

Exploration and communication indeed seem to be two aspects of the same course of action. By communicating with other humans, we became conscious of ourselves. Doing so with nonterrestrial intelligent beings will bring us a different image of ourselves and will relativize human existence.

A space probe has one obvious advantage over exploration with a telescope: it goes *in situ* to gather information, and if necessary it adapts its observing program to the specific conditions it meets and to preliminary results. Information does not have to be exchanged with the pilot station. This aspect would be particularly important in contacts with distant civilizations. We can imagine a situation in which an automatic probe makes the first contact, elaborates a common language, then sends us any useful information. This method would save considerable time in comparison with direct communication across hundreds or thousands of light-years.

If there are other civilizations in our Galaxy, why haven't they invaded it yet with exploratory probes and colonizing spaceships?

Our noisy society (especially in the range of radio waves) should attract them to the solar system. The obvious absence in our solar system of a swarm of probes of various origins seeking contact is the crucial point of the present debate among astronomers. Some assert that this absence is proof that we are alone; others respond that, statistically, "they" have not had the time to visit us.

Although curiosity, and therefore exploration, are values proper to all forms of intelligence as we imagine it, obviously expansion and colonization are very differently motivated. In an earlier chapter I reviewed the reasons — historical, economic, political, ecological — that could inspire our civilization to attempt colonization. In the absence of signs of extraterrestrial life, I shall now examine the reasons why "they" might not show themselves, if "they" exist.

It seems just as extreme to state that we are alone because if they existed they would already have colonized the whole Galaxy, including the solar system, as to assert that life, and then intelligent life, constitutes the inevitable path of evolution in any location in the Galaxy. There is no major reason why we should be surprised that we don't see any extraterrestrial colonies within the range of our telescopes.

What is more disturbing is the absence of exploratory probes, signal-carrying electromagnetic waves, interstellar beacons, or other signs of intelligent activity in the universe. But absence of proof is not proof of absence. I shall give four possible explanations for this state of affairs, beginning with the more likely ones.

We are alone. We are alone, and we have always been alone, because life is a unique phenomenon. This is one of the possible conclusions of this book, the one supported by biologists against astronomers. Were life unique, this would not be in apparent contradiction with the immensity of the universe, as we shall see below. But evolution toward greater complexity is characteristic not only of life on Earth but of all matter in the universe. Would that not be a good reason to extend the "principle of mediocrity" (according to which our planet is just ordinary) to life? To extrapolate this principle is to push the Copernican attitude to its limits.

Another possible reason for our solitude is that other civilizations may have disappeared because of internal crises. The nucleus of the Galaxy could also periodically go through the activity state that characterizes Seyfert galaxies,[1] producing intense radiation (particularly x rays) that would greatly disturb any forms of galactic life. Finally, mass extinction of living species on Earth could be the result of a phenomenon universal enough to affect all life in the Galaxy.

We did not recognize them for what they were. It is possible that we have detected signals coming from other civilizations but that we interpreted them in terms of known natural phenomena. A researcher's normal reaction upon encountering a new observation is to check his repertoire of known and more or less understood phenomena before looking for an exotic interpretation.

The luminosity of a considerable number of stars in the Galaxy varies with time. This variation can be erratic, periodic, or quasi-periodic, and can be on a time scale from a fraction of a second (pulsars) to several years (Mira-type stars). We could very easily imagine that some of these variable stars are in fact artificial sources of light and that the variation is a message. It is perfectly possible that astronomers have already observed artificial signals, but that they simply classified them in their catalogs of variable stars without looking for the origin of the variability.

It would not be the first time that an astronomical phenomenon has been incorrectly interpreted. For a long time, the prototype of a class of quasars, the Lacertids, was believed to be a star, cataloged under the name BL Lacerta. In defense of astronomers, I must add that it is very difficult, through physical theories, to account for phenomena that are variable with time, especially when they are not periodic. The evolution of a star and its atmosphere have to be followed by computer, but the phenomena under consideration are so complex that they quickly saturate the memory of the most powerful computers if we want to take all the star's

1. Seyfert galaxies closely resemble quasars. They have a very brilliant nucleus where the gas undergoes very violent turbulent motion, probably because it is collapsing into a central black hole.

(Carillon, Paris, for Sempé)

"But think for a moment, Armand. If there really are beings with superior intelligence, why would they send signals to *you*?"

properties into account. The execution time of computer programs also becomes impossibly long, because events which last millions of years have to be simulated using short time steps.

The problem lies in identifying what distinguishes a signal of artificial origin from a signal produced naturally without the intervention of intelligence. Most natural emissions produce radiation spread over a wide range of wavelengths; this is especially the case for stars. The criterion adopted by radio astronomers is that an artificial signal is emitted at a single wavelength. But this is not a sufficient condition, since hydrogen and interstellar molecules also emit in that way.

We have not yet looked in the right place at the right time with the right instruments. The search for extraterrestrial civilizations has just begun, and we are still far from having explored the whole Galaxy. Furthermore, the observable universe is inexhaustible: each new generation of astronomical instruments brings a harvest of important discoveries at a pace that has been accelerating since the 1960s, thus revealing the complexity and richness of the universe. The fact that several thousand stars have been sporadically observed for the past twenty years with no positive results suggests, at most, that the number of technological civilizations in the Galaxy is probably low — on the condition that they are not hiding. This leads us to another possibility.

They do not show themselves. To examine the possible reasons for such an attitude, I shall once again adopt an anthropomorphic point of view and ascribe human behavior to extraterrestrial civilizations.

- The Contemplation Hypothesis. They look to an inner world from which we are absent. They are not motivated to explore and communicate, because they have nothing left to learn or because knowledge does not interest them.
- The Ecological Hypothesis. Despite their interest in exploration and communication, they do not want to modify their environment or to waste energy by sending probes and radio waves — for perfectly understandable reasons. From a

slightly different point of view, launching automatic and intelligent space probes might pose risks if they manage to escape their creators' control.

- The Hypothesis of the Mental Horizon. From their point of view we are a primitive civilization whose mental horizon is very limited and therefore of little interest. The physics of their communication techniques is beyond our grasp. We are like a primitive people living in a jungle, unaware of modern civilization and incapable of intercepting radio broadcasts. But, to pursue the analogy, why don't we see other demonstrations of their existence as evident as are airplanes to primitive people?
- The Immaturity Hypothesis. Too early a contact would change us into information consumers. It would prevent us from independently developing a unique experience of evolved life, an experience that would otherwise enrich the galactic culture.
- The Zoo Hypothesis. We are a primitive civilization living on a galactic reservation. They do not enter for the same reasons that we keep out of nature preserves on Earth. In this hypothesis, we could imagine a science fiction scenario in which UFOs are gamekeepers who show themselves by accident.
- The Distrust Hypothesis. Our technological progress and our tormented world history make us suspect. In this case, and remaining in the realm of science fiction, UFOs are guards who survey our activities. Certainly our exuberance in certain fields — radio broadcasts in all directions, erratic nuclear explosions, various kinds of pollution — all suggest handling with care.

All these hypotheses, so successful in science fiction, suffer from the same serious defect: to be correct, they should be applicable to all the civilizations in the Galaxy. But these hypotheses, extrapolated from a very resolutely anthropomorphic attitude, do not even apply to our own civilization. We could imagine other, non-anthropomorphic hypotheses to explain why they all hide, but the defect that we are an exception would still be there. Thus, the simple fact that we are trying to explore and communicate invalidates beforehand any hypothesis that justifies the unwillingness of an extraterrestrial civilization to show itself. Otherwise, we are exceptional and it is therefore useless to try to find others like ourselves.

However, these hypotheses could be applicable to only a certain proportion of the civilizations in the Galaxy, thereby reducing the number of detectable ones who are more willing to communicate with us.

I have just given a fairly wide range of reasons for our not yet noticing them. Most of these reasons suppose their existence. One point must still be cleared up: if they exist and if they are inclined to explore the Galaxy with automatic or manned probes, why have their probes not come to Earth to contact us? There are several explanations.

They have already been on Earth. This would explain a certain number of architectural feats and evidence of scientific and technical skills found in the remains left by past civilizations whose level of development we underestimate. But then we would have to explain why they did not stay. Or else, they came at a time when man did not yet exist. But in that case, why did they not leave a probe in orbit around the Earth? They could then monitor the evolution of the planet, which according to their own experience would be a favorable site for the beginning of a new civilization.

They are established somewhere else in our solar system and are watching us from a distance. It is not necessary to go to the center of the solar system to find the basic elements for a colony. The fragmentation of the asteroid belt could be the result of mining by colonizing civilizations. (Table 10 makes an inventory of the resources of our solar system that could be useful for colonizing civilizations.) But in this case, why haven't we detected them yet?

They don't want to visit us, for one or several of the reasons given above which prevent them from revealing their existence to us.

The solar system's environment is hostile to their probes. Since we do not occupy a privileged place in the universe (the Copernican principle), this could mean that a good part of the universe is hostile to them and that they have not mastered the problems of interstellar travel.

Table 10
Distribution of the elements in our solar system and their use for interstellar colonizers. (After Stephenson 1979.)

	Atomic weight		
	light	medium	heavy
Elements	hydrogen helium lithium beryllium	oxygen carbon nitrogen	iron nickel copper silver
Purpose	fuel	life	structure
Use	single pass	recycled	static
Losses	used up	flushing	erosion, wear
Needs	important	medium	small (repairs, growth)
Abundance	95%	1%	traces
Distribution	outer planets, crust of Earth, ices of outer moons	outer planets, Venus, Earth, asteroids, carb. chondrites	cores of planets, asteroids, inner planets

They have not had the time to come to Earth. This is the simplest explanation and the one that has attracted the most research by astronomers over the past few years. The various models for the colonization of the Galaxy use our human experience of colonizing Earth as a starting point, then extrapolate to interstellar travel by adapting them to the special conditions in the Galaxy. There are many unknown parameters, including the speed of the colonizing spaceships, the length of pauses between waves of colonization, the colonizers' internal rate of population growth, and the distance between colonizable sites.

The most optimistic astronomers claim that to colonize the Galaxy a society would need from 5 to 60 million years. However, a society with no demographic growth would need an infinite length of time to do so.

Using automatic probes changes the situation. The colonization time would depend mainly on the technological development of the colonizing society. Exploratory probes traveling at 30,000 km/sec (one-tenth the speed of light) should be able to explore the Galaxy in 4 million years. However, these calculations do not take into consideration the delays that would be caused by the probes' exploiting each new discovery before leaving.

9

UFOs and Paleovisits

Vision is the art of seeing invisible things.
—Jonathan Swift, "Thoughts on various moral and amusing subjects"

"Unidentified flying objects" are not always objects. It is better to talk about *unidentified aerospatial phenomena*. The Groupe d'études des phénomènes aérospatiaux non identifiés (GEPAN) gives the following definition: "elusive, generally bright phenomena which are located in the atmosphere or on the ground and whose nature is not known or recognized by the people who observe them."

These phenomena, although exceptional, are not rare. During the 13 years I worked at the Observatoire de Besançon, not a month went by without several phone calls to the switchboard concerning UFO appearances in the area and asking for an astronomer's opinion. Claude Poher's statistical study, which I will discuss below, is based on a file of 825 UFO sightings in the entire world, 220 of them in France.

In 2/3 or 3/4 of the cases, ordinary events were seen under extraordinary conditions: airplanes in strange circumstances, meteorological or atmospheric high-altitude (30 – 50 kilometers) balloons lit up by the Sun when it is already night on the ground, strangely shaped clouds lit by the setting Sun, artificial satellites, projectors or spotlights reflecting on clouds, flashes of lightning, aurora borealis, etc. Somes of the sightings proved to be astronomical phenomena: meteorites, fireballs, stars (Capella, Sirius), or planets (Venus, Jupiter).

In addition to cases unsolved because of scanty evidence, there are always a few troubling cases that are not accounted for by known natural phenomena. These unexplained leftovers are the ones that deserve scientific attention. They may be meteorological or astronomical phenomena that are still unknown. There is no need to start explaining them by invoking extraterrestrial civilizations. It is above all to forestall the reader's questions or eventual criticisms that I take up this subject.

This phenomenon, elusive by its very nature, is very difficult to study scientifically. Evidence is often very vague, and sometimes the same event is described in contradictory terms by several witnesses. Witnesses tend to attribute speed or size to UFOs that they saw at night, with no reference points. At night it is generally not possible to estimate anything other than an object's shape, its color (although our perception of colors is reduced in the dark), the direction in which it was moving, and the length of time of the sighting. These are the questions astronomers ask when people call. Some people exaggerate so that they will be taken seriously. Traces are sometimes found on the site where the UFO appeared, but these may not have any connection with the phenomenon.

Psychological and sociological factors also complicate the scientific study of UFOs. The attraction of the extraordinary and the supernatural convinces witnesses that they saw an exceptional phenomenon. Hoping to meet visitors from outer space also alters a witness' objectivity. A number of books and films about contacts of the "third kind" have introduced UFOs into contemporary mythology. A few years ago, a young man monopolized the attention of astronomers at my observatory with a slide on which a large, black spot was discovered after the film was developed. He thought it was a UFO. It was probably a bird. A local newspaper did not hesitate to buy his picture.

According to the psychoanalyst Carl Jung, UFOs could be the resurgence, in a pathological form, of our inclination for the irrational, but muzzled by a conscious desire for order and rationality. The unconscious appropriates a natural phenomenon seen in the sky and deforms it by giving it another meaning, in particular by exaggerating its properties. And indeed, most UFOs turn out to have natural explanations. This does not mean that those who witness UFOs are visionary or strange. Gallup polls taken in 1966 and 1973 showed they do not differ from the rest of the population by religion, profession, political opinion, or level of education.

We know the opinion of the average Frenchman about UFOs from a SOFRES poll conducted in June 1985 on a national sample of 1,009 people aged 18 years or over. To the question "Do you personally believe that extraterrestrials have already been on Earth?" 21 percent answered Yes, 68 percent No, and 11 percent were without an opinion. There was no significant variation between the sexes, but young people believed more — 34 percent of those between 18 and 24 answered affirmatively. This state of mind inspires witnesses of strange phenomena to interpret them in terms of UFOs.

The public's gullibility and the media's greed for sensationalism have also favored mystification. Several photographic documents purporting to show UFOs have turned out to be forgeries. One of the phone calls I received was in fact a hoax by a local radio station. Such factors tend to make scientists skeptical when UFO witnesses call them. Moreover, mystifications discredit other accounts.

At the request of the US Air Force, for two years a group of scientists at the University of Colorado analyzed UFO accounts made in the United States. The results of this Scientific Study of UFOs, known as the Condon Report, were published in 1968. The report's authors recommended no longer studying the UFO phenomenon by the same methods — analysis of witnesses' accounts — because it did not make it possible to determine whether the UFO phenomenon was the result of (1) incorrect identification, incomplete accounts, and mystifications or (2) extraordinary events. Witnesses' accounts are equally compatible with either hypothesis.

The authors of the Condon Report were not unanimous about the conclusions; one of them, David Saunders, even had to resign. Nor were astronomers unanimous about them. Peter Sturrock of Stanford University thought that the conclusions contradicted the body of the report, which noted unexplainable accounts. J. Allen Hynek, a major figure in this field, was of the opinion that the study was incomplete. For example, it did not emphasize the hundreds of accounts of magnetic or electrical effects linked to UFOs. He felt that the study had not been adequately funded, and he thought such work should be continued. According to Hynek, accounts should be classed by two criteria: the degree of strangeness and the degree of the witnesses' credibility.

In France, in 1976, Claude Poher analyzed nearly a thousand UFO reports from all over the world. He mainly tried to classify the accounts according to very detailed criteria, but he also used Hynek's two criteria, strangeness and credibility, to which he added a third: the quantity of information gathered. He studied the distribution of the various classes of reports as a function of different variables (length of the phenomenon, hour of the apparition, geographic location), but he was unable to identify any consistent patterns.

In 1977, the Centre National d'Etudes Spatiales created GEPAN to scientifically study UFOs. The analysts first gathered all pertinent information on the subject, especially from the police; then they began to analyze the results of inquiries and statistical studies of witnesses' accounts. Along with this, they developed a research method that considered the psychological and sociological status of the witnesses, the physical conditions during the observation, and the socio-cultural environment of the account. By placing the event in its overall context, the GEPAN analysts hoped to determine the exact nature of the stimuli that had produced the accounts. They also developed research in fluid mechanics and in the psychology of perception.

As the reactions to the Condon Report show, scientists are far from having the same opinion about how much importance should be accorded to UFOs. Edward Condon thinks that it is not necessary to accept the hypothesis of extraterrestrial civilizations to explain UFOs. Others, including Carl Sagan, declare categorically that our little planet does not deserve so much attention from possible extraterrestrial civilizations. In Sagan's opinion, public interest in UFOs comes not so much from scientific curiosity as from unsatisfied religious needs — UFOs have replaced the gods that science deposed.

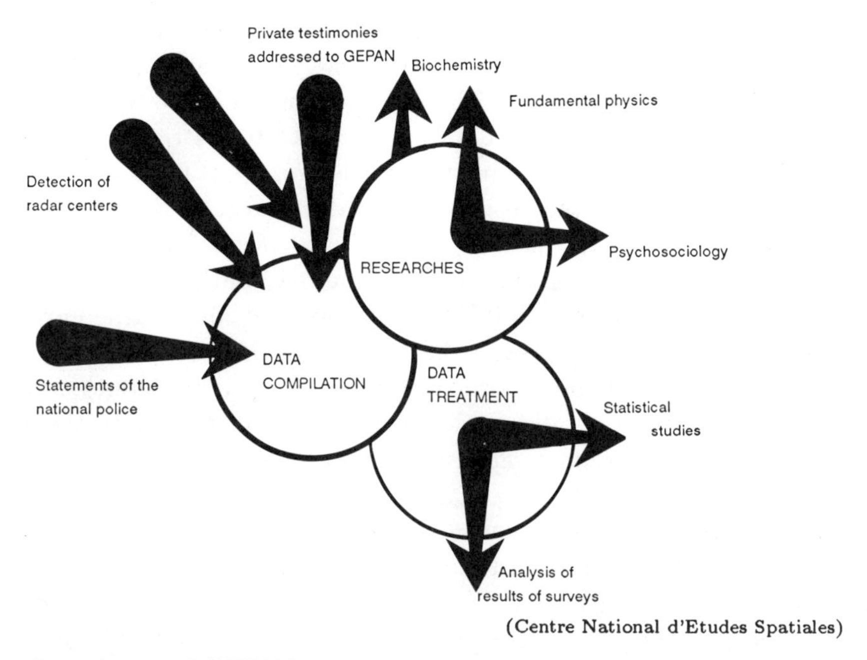

(Centre National d'Etudes Spatiales)

Organigram of GEPAN.

An inquiry made in 1975 and 1976 by Sturrock among members of the American Astronomical Society revealed that the subject is delicate, that most astronomers who answered the questionnaire were of the opinion that the subject deserves scientific study, that most of them were ready to participate in learning about the nature of UFOs but did not know how to proceed, and that several percent of them could furnish accounts similar to those coming from the same proportion of the public.

In my opinion, the UFO phenomenon deserves scientific study, but will probably not prove that extraterrestrial visitors have been here. My attitude may seem paradoxical, since I devote several chapters to hypothetical extraterrestrial civilizations; however, the scientific method requires looking for an answer among known natural phenomena before going to less likely hypotheses.

My attitude is the same about black holes, often invoked by American colleagues to explain any excess of matter revealed through secondary effects but invisible in photographs of the sky. Just as it is too easy to use black holes as an excuse when conventional explanations fail, it is too easy to explain phenomena that don't yield to simple interpretation by calling them UFOs.

Not finding convincing proof of recent extraterrestrial visits on Earth, several writers have dug into our historic past to seek traces of contact with possible visitors from outer space.

The main argument used in favor of "paleovisits" is that past civilizations were too primitive to be the only authors of certain achievements or knowledge. The astronomical and mathematical sciences of the Egyptians, the Maya, and the Sumerians, and the Dogons' astronomy, were supposedly inspired by interstellar travelers. The Egyptians would have been unable to build their pyramids without outside help; the stones of Baalbek were impossible to move with the means available at the time. Such interpretations insult the civilizations concerned, which did not reveal to posterity the sources of their knowledge or the extent of their technical competence.

A second category of alleged proofs link certain architectural realizations to extraterrestrial visitors. The figures of Nazca, Peru, visible only from the air, were for communicating with extraterrestrials. The terraces of Baalbek were landing strips. The mystery of the Easter Island statues is also connected to extraterrestrials. These are all useful alibis for our ignorance.

Finally, passages in ancient texts seem to mention extraterrestrial visitors, or at least phenomena that might be ascribed to them. According to the Soviet astronomer M. M. Agrest, Sodom and Gomorrah were destroyed by a thermonuclear explosion. He also thinks that the catastrophe of Tunguska in 1908 was caused by the crash of a spaceship and that the sacred Hindu texts Ramayana and Mahu Bhratra contain descriptions of extraterrestrial visitors. The sensationalist German writer Erich von Däniken interprets some of the Maya bas-relief figures in Palenque, Mexico, to represent astronauts with spacesuits and oxygen masks. He also claims that the Sphinx at Giza represents an extraterrestrial.

In their book *The Morning of the Magicians*, Louis Pauwells and Jacques Bergier devote several chapters to possible visits by extraterrestrials in the past, and conclude "we do not refuse to suppose visits from inhabitants coming from outside."

All these sensational revelations by writers who may be good journalists but are not scientists should be taken with a lot of skepticism, if not complete disbelief. In my opinion, only one case deserves detailed examination: Dogon astronomy.

The Dogons are a nomadic people of Mali whose knowledge of astronomy is remarkable. For the Dogons, the universe is infinite but measurable, filled with spiral stellar worlds, on one of which lies the Sun. The universe is divided into two systems. Most of the stars in the celestial sphere form the external system, which has little influence on earthly life. The internal system participates directly in life and is made up of Orion, Sirius, Gamma Canis Minor, Procyon, the Pleiades, etc. Among these stars and stellar clusters, Sirius plays a fundamental role. The star Sigi Tolo (Sirius) has a small companion, Po Tolo (Sirius B, discovered by Alvin Clark in 1862), which circles Sirius in 50 years (the period is in fact 49.9 years). Po Tolo is the smallest and heaviest star (it is indeed a white dwarf, whose remarkable properties — its extreme density in particular — were learned only in 1915). A second companion, Emme Tolo — 4 times lighter and a bit bigger than Po Tolo — also circles Sirius with the same period but with a slightly larger trajectory. This latter prediction contradicts Kepler's law of areas. As for the star, astronomers have searched for it unsuccessfully.

The Dogons do not seem to have had sophisticated methods of observation, since they did not know of the existence of planets beyond Saturn. But even if they had the means of observing the system of Sirius and establishing its periods, predicting the high density of the white dwarf required advanced scientific knowledge, unless it was just luck. Interference by a latter-day astronomer seems unlikely, because this astronomical knowledge is well integrated in Dogon culture and because some of the theories (such as Sirius' third companion, or the idea that the second companion becomes brighter when it approaches Sirius) are unknown to astronomers of our own time.

The source of the Dogons' astronomical knowledge deserves further study. The breadth of their mythical vision of the universe should be specified, and analogies with mythologies and knowledge of other peoples in the area should be looked for.[1] However, as in the case of UFOs, I refuse to invoke extraterrestrial intervention even though I cannot find other satisfying explanations.

1. The knowledge extant in the sixteenth century at the Muslim University in Timbuktu may be particularly relevant.

10

Strategies of Searching for Extraterrestrial Civilizations

Nature gave us two ears and only one tongue so that we may listen more and talk less.
—Zenon of Elea

In view of the many uncertainties I have discussed, there is only one possible course of action for astronomers: to start listening to the universe in the hope of detecting signals of artificial origin.

But what kind of signals should we look for? If we examine the particular case of life on Earth, we notice that it has caused a relatively modest imbalance in its environment — except in the range of centimetric and millimetric radio waves, where it is particularly noisy. Should this lead us to deduce that the most easily detected form of life is an extraterrestrial civilization that emits radio signals? One important area of bioastronomical research is based on this hypothesis. Astronomers are betting on the existence of extraterrestrial civilizations that transmit radio signals.

Without actually describing them, the Soviet astronomer Nikolai S. Kardashev classes these hypothetical civilizations in three categories according to their ability to master energy. Type I civilizations have domesticated the same quantity of energy as we have, Type II civilizations have mastered the energy of their star, and Type III civilizations have mastered the energy of their galaxy.

Such a classification is useful for designing a search strategy. Our civilization is easily detectable by its radio emissions. A Type II civilization could have surrounded its star with a "Dyson sphere" — a thin shell of planetary material as far away from its star as Jupiter is from the Sun. It would be detectable only by its energy losses in the infrared range, unless it used a part of this huge amount of energy to transmit interstellar signals. The energy available to Type III civilizations should also be conspicuous, even on intergalactic scales, if it is used for signaling.

However, this research faces many problems — both conceptual and practical. First, our physical theories are limited. Radio waves, on which most research concentrates, might not be the best way of communicating. More efficient methods that we are not yet aware of are perhaps being used for communication among extraterrestrial civilizations.

Even if they are using radio waves, numerous unknowns remain: the signal frequency, the way the wavelength is modulated, the orientation of the receiver (or, conversely, the location of the transmitter), the broadcasting schedule, and so on. Astronomers have been ingenious in optimizing all these parameters and thus reducing the size of the "cosmic haystack" in which the artificial signals are buried. They have acknowledged the necessity of developing new receivers which chop up the incoming radio radiation into millions of channels of very narrow bandwidth and which are capable of instantly analyzing the signals in each channel.

Even if one day we manage to receive a message, the problem of decoding it remains. What could the grammar and semantics of an extraterrestrial language possibly be? The concrete search for signals, as well as the study of the concept of an extraterrestrial civilization and its consequences, forces us to place notions of which we have only one example into a relative context. To add to the difficulties, such research must be financed despite the skepticism of those who control the funding of science.

Several options exist for detecting extraterrestrial civilizations. We could attempt to intercept signals coming from an artificial source. We could also transmit signals, in all directions or in specific ones, hoping that they will be intercepted and answered. We could send automatic or manned exploratory probes. The best strategy is to adopt the most economical and efficient method.

The simplest and cheapest method, but maybe not the most efficient one, is the interception of signals emitted by an extraterrestrial civilization. These signals might be emitted for several reasons: to make the civilization's presence known to us (by an interstellar beacon, for example), or to converse with a probe or another civilization with which contact has already been established,or simply as "noise" accompanying the civilization's usual activities. This is the strategy adopted by almost all the astronomers involved in such research.

We have been capable of communicating over great distances for only a few decades. It is therefore reasonable to limit ourselves to listening for more advanced civilizations who have greater energy resources and more sophisticated technology to devote to their broadcasts. Furthermore, it is not really certain that we have something interesting to transmit. But even if this strategy may seem to be the best for a certain period of time, it will have to evolve, for a very simple reason: if all civilizations adopted this strategy, none would transmit anything!

Even if we don't transmit messages, extraterrestrial civilizations may be listening to us and would in fact intercept many signals coming from our planet. But any that have done so must be rather close by, as we have been emitting radar and telecommunication signals, and radio and television broadcasts, for only a short time.

Simultaneous transmission and interception of signals solves the problem raised above and permits more efficient short-term "semi-contacts" rather than dialogues where response times are very long. In this case each civilization transmits information spontaneously, without trying for dialogue, while at the same time it listens to others. The energy needed to send "audible" periodic messages could be drawn from hydroelectric or nuclear power plants during periods of low domestic demand.

Such one-way communication — a double monologue — may appear uninteresting. It is nevertheless a common way of communicating across the centuries, if not across space. Consider world history. The civilizations of the past — Greek, Roman, Arab, etc. — have communicated to us part of their knowledge and culture, without any hope of benefiting from ours.

The third option is exploring the Galaxy with automatic or manned space probes. The efficiency of automatic probes in exploring the solar system does not have to be proved. If we had sent messages rather than the Viking probes to Mars, we would still be asking ourselves if there is life on Mars. However, the cost of space research is very high. Money could perhaps be found through public subscription or by selling products of space exploration, such as cassettes taped on Mars, where the acoustics would be quite extraordinary. But the time needed for an interstellar journey makes it a very long-term project and a commitment for which great political and economic stability is necessary. It is a method for the future — one that we will use in several hundred years.

The whole universe is filled with natural signals transmitted at all possible frequencies by stars, the Galaxy, and the interstellar medium. Thus an artificial signal, to be intercepted and understood among all this noise, must follow the principle of anticryptography. A society wishing to enter into contact with other civilizations should optimize the chances for its signals to be detected and decoded. This consists in putting ourselves in the observer's place, whether he is looking for artificial signals or simply doing classical astrophysics. We then ask ourselves what kind of receiver the observer uses most often, at what frequency and modulation he will probably set his receiver, how he will direct it, what his listening hours will be, and so on. Finally, we must broadcast a signal whose contents will be easy to decode and interpret.

An observer searching for artificial signals should reason in a similar way. He should ask himself what are the characteristics of an artificial signal that would be easy to detect among the galactic noise. Most astronomers agree on one property of such a signal: it should be broadcast in a very narrow bandwidth.

We perceive the universe by the radiations it transmits, or at least by those that we know how to intercept through our senses or through our receiving devices. The study of these radiations is the object of astronomy. In the current state of our knowledge, electromagnetic waves, which are widespread throughout the universe, are the most interesting type of radiation for communication. Gravitational waves, predicted by Einstein's Theory of General Relativity, are extremely difficult to detect. Sound waves travel only in the atmosphere, and do so very slowly. Cosmic rays are composed of charged particles which generally come from supernova explosions in the Galaxy's disk. Since these particles are charged, they are deflected by the solar wind and by magnetic fields, so they could not be used for sending messages to a precise target.

Electromagnetic waves are emitted at all frequencies: low frequencies correspond to radio waves; infrared and visible light are at medium frequencies; ultraviolet radiation, x rays, and gamma rays are at very high frequencies. Each domain of frequencies needs its own special receiver. Some of them must be placed aboard balloons or satellites, because the Earth's atmosphere absorbs ultraviolet and far infrared radiation, x rays, and gamma rays.

All these waves travel at the same speed, that of light — 300,000 km/sec in a vacuum. This is the fastest that information can travel. All known matter in the universe radiates electromagnetic waves naturally in the whole range of frequencies. We are therefore justified in expecting that an extraterrestrial civilization would use these waves, which are so common in the universe, to communicate.

The probability of intercepting electromagnetic waves carrying messages not meant for us is probably quite small if we suppose that physics offers infinitely more communication possibilities than we know about. In that case, electromagnetic-wave communication would be used only by the limited number of civilizations at about our own level of development. More advanced civilizations would certainly prefer a more efficient and economical method — one that we don't know about yet because it is part of a physics and a technology of which we know nothing.

If electromagnetic waves are indeed transmitted artificially and are really being used for interstellar communication, we should be able to distinguish them from those that are emitted naturally. A signal transmitted as a wave could be divided this way:

Signal = Frequency + Modulation + Code

or else this way:

Signal = Physics + Technology + Language.

The signal is carried by a wave emitted by a physical mechanism in a certain frequency range. This wave can be artificially modulated by some technique, and the rhythm of modulation constitutes the message. If the modulation is random there is neither code nor message — the wave is just natural "noise," modulated by a natural phenomenon. It is thus mainly the way the wave is modulated[1] that allows the distinction to be made between natural and artificial signals. The frequency of emission can also reveal an artificial signal if there is no physical mechanism that allows natural emissions at that frequency. Another possibility is broadcasting at a frequency where nature does not spontaneously emit radiation, for example a natural frequency multiplied by an irrational number such as π or $\sqrt{2}$. Indeed, no natural physical process can displace the frequency of a wave in this way. The signal then becomes

Signal = Frequency × Irrational Number.

This possibility is valid for an interstellar beacon intended to signal the presence of intelligence, but it does not allow communication. Such signals would thus show us that we are not alone, but no more. One could, of course, include a message in the signal at a later stage.

Let us come back to the first form of transmission suggested: a wave emitted at a certain frequency, with the message contained in the temporal modulation of the wave. The reception of such a signal depends on several parameters. First, the antenna or telescope — the receiver in general — must be oriented toward the source of the signal. The receiver must then be turned on at the right moment, and tuned to the right frequency. Finally, it must be set to detect the correct form of modulation. Each of these different variables has an infinite choice of values. We must optimize this choice in order to have a chance of detecting an artificial signal, or of being detected if we are the ones broadcasting.

1. There are four ways of modulating an electromagnetic wave: in phase, in polarization, in amplitude, and in frequency. The last two methods are commonly used for radio and television broadcasts. Circular polarization eliminates interference of terrestrial origin.

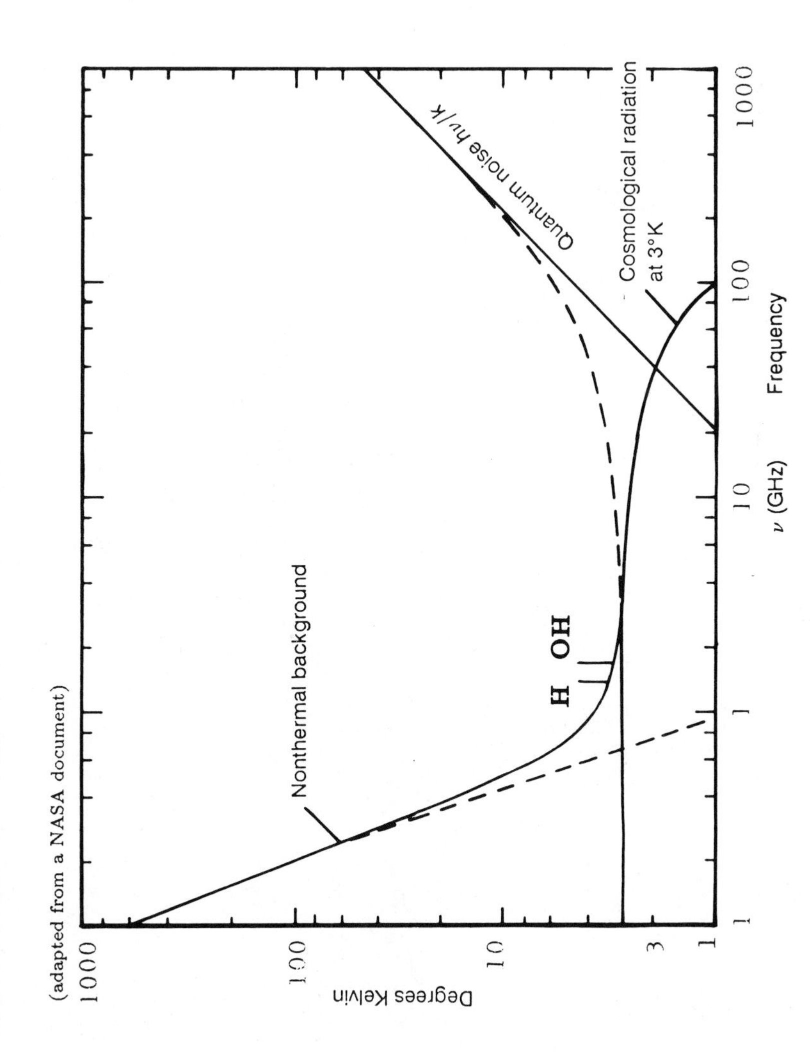

The microwave window. Outside the plane of the Galaxy, the nonthermal background noise limits the sensitivity of receivers at low frequencies and quantum noise limits it at high frequencies.

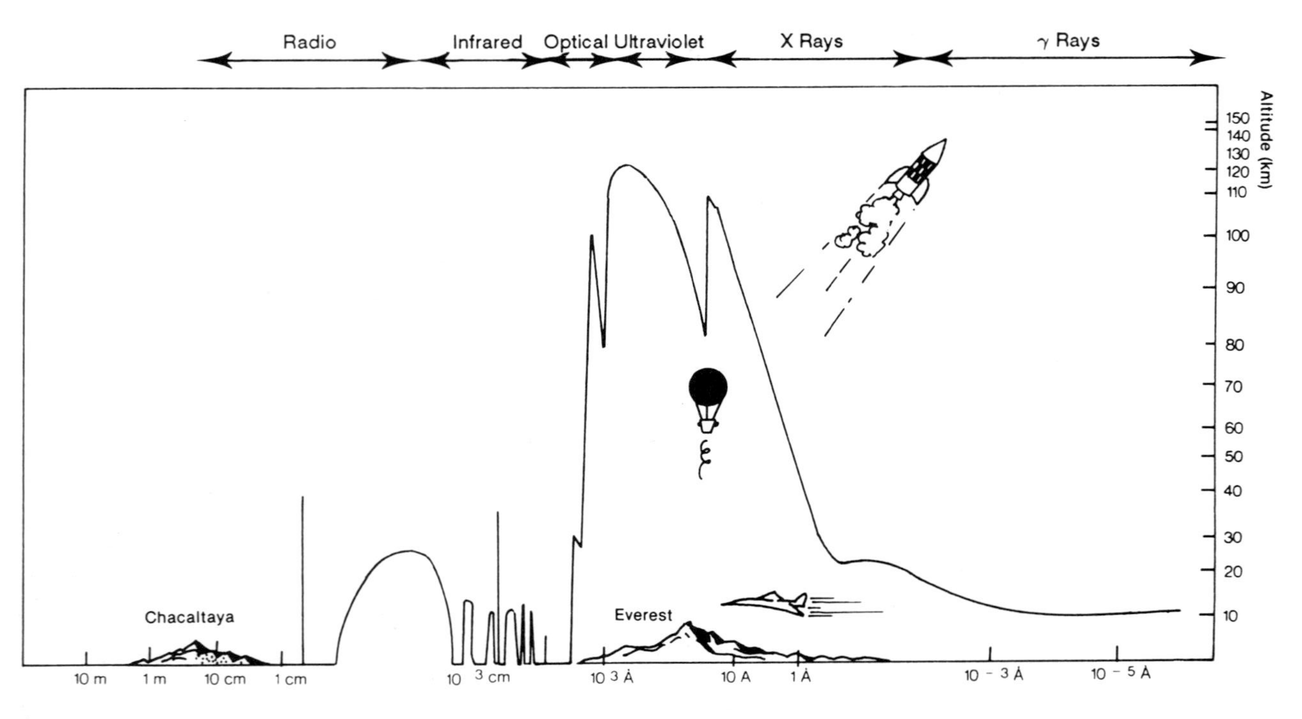

(Adapted from "The Galaxy, the extragalactic universe," ed. Bureau des Longitude, Paris)

Transparence of the terrestrial atmosphere. This diagram shows how high we have to go in order to be able to collect radiation coming from the universe. Some infrared and gamma frequencies are accessible by airplane.

There are an infinite number of possible frequencies for transmitting signals. However, our ability to receive signals from outer space is limited by the Earth's atmosphere, which is transparent in only two ranges of frequencies: optical and radio. It is therefore appropriate to start with frequencies in one of these two areas.

The visible domain is not very favorable, because light rays are diffused and absorbed by interstellar dust and because optical telescopes "see" less far into the universe and with less resolution than arrays of radio telescopes.

Radio waves offer a number of advantages. The background noise in the Galaxy is minimal in the domain of radio waves. Since we will have to detect a signal whose power will be attenuated by the distance it travels, this is an important advantage. There are "key" ranges of radio frequencies in which radio astronomers often observe for astrophysical purposes. They have nicknamed "the water hole" a frequency band between 1,400 and 1,800 megahertz (i.e., a wavelength of 18 to 21 centimeters) of natural emissions of the components of water: hydrogen and the hydroxyl radical (OH). Hydrogen is the most abundant element in the universe, and neutral hydrogen atoms radiate spontaneously at the wavelength of 21 centimeters. Any civilization that has started exploring the universe must have detected this universal radiation. Furthermore, radio receivers have several advantages over optical telescopes: their collecting area is larger, they are cheaper, they last longer, Doppler effects are smaller and more easily compensated for, and observing does not require good weather. From an Earthly point of view, the radio frequency band is the best choice.

The Galaxy's natural radiation is very broad-band, sometimes spread over the whole electromagnetic spectrum. The Crab pulsar is an example: it emits mainly in the radio domain, but also in the visible and x-ray ranges. Therefore, in order to stand out against the continuous background noise, an artificial signal should be transmitted in the narrowest possible band. The same amount of energy emitted as a signal in a narrow band will carry further than it would in a wide one.

A narrow-band signal would, however, be difficult to detect if we did not know the frequency in advance. The reason is simple:

if our receiver has a broad band, the signal would be spread over the whole band and would be undetectable unless it were very powerful. If our receiver has a narrow band, we can observe only a small section of the radio domain at a time (unless the receiver has very many channels), and the message might not be noticed. Another drawback of a narrow-band signal is that the narrower the emission frequency, the more limited the modulation (and thus the message). Finally, detecting an artificial signal by accident is not very likely. Astronomers usually reject all parasitic noises that hinder their observations without bothering to identify them.

One strategy of listening and transmitting is to choose one of the "key" frequencies mentioned above, or its immediate neighborhood. But anyone who wants to listen or transmit at a particular frequency must first solve a serious problem: a message transmitted at a given frequency will not be intercepted at the same frequency by a receiver moving relative to the transmitter, since the wave frequency is shifted by the Doppler effect. The solution that best suits the anticryptography principle is to listen or transmit at a frequency that takes into consideration everyone's motion relative to the whole Galaxy. This does not entirely solve the problem, because we know our own motion with an uncertainty of 10 percent. This uncertainty is important when the signal is at a frequency multiplied by an irrational number.

One strategy would be to transmit easy-to-detect but messageless signals over several frequencies, and to broadcast the message itself over the central frequency. A listener would be intrigued by a series of signals that get closer together and more intense as he sweeps over the range of frequencies, then weaker and further apart. If he is clever enough to examine what is happening in the center of this system of symmetrical signals, he will probably detect a meaningful signal.

Technical reasons could also influence the choice of frequency. When low-frequency radio waves travel through the interstellar medium, their bandwidth is broadened. On the other hand, low-frequency waves travel much further than high-frequency ones.[2]

2. The range of an electromagnetic wave of given energy is inversely proportional to its frequency.

The choice of targets to listen to is another dilemma for radio astronomers. Common sense tells us to observe all the stars like the Sun — G-type stars — in the immediate neighborhood of the solar system. This is, in fact, where the search for artificial signals started. However, if the civilizations that we are looking for have spread throughout the Galaxy, we could find them around any star. Some astronomers prefer to observe systematically, sweeping the whole sky with their antenna. The search for stars identical to the Sun in the Smithsonian Astrophysical Observatory catalog, which contains information on 260,000 stars, has yielded meager results. Among the 50,000 G-type stars listed, very few have been well studied. One may nevertheless select as targets about a thousand stars toward the galactic center and in the opposite direction.

Another option is that of observing all the abnormal sources of radiation which have been catalogued in the course of classical observation. The optical identification of cosmic radio sources is an important part of a radio astronomer's job. Depending on which wavelength was used for making radio maps of the sky, the radio sources could be gaseous nebulas, supernova remnants, galaxies, or quasars. When these sources turn out to be distant stars, they are often good candidates for artificial-signal research. Generally the radio activity of stars is quite moderate compared with other radio sources. The most intense radio source in our sky is the Sun, but this is due to its closeness. One abnormal stellar radio source has been reported, and I will discuss it in the next chapter.

Some stars have excessive infrared radiation because they are surrounded by a dust cloud. But this same radiation could be the result of interstellar engineering. Freeman Dyson thinks that at a stage of advanced technological evolution we will live inside a hollow sphere circling the Sun and made of matter from Mars and the asteroids. This "Dyson sphere" would intercept all the Sun's energy and re-emit wasted heat in the infrared range.

We could try to optimize the chances of detecting signals by pointing the receiver toward a region where stars have accumulated: the center of the Galaxy (several billion stars), globular clusters (one million stars), or another galaxy (hundreds of billions of stars). Unfortunately, the distances to those kinds of objects are astronomical — 6,000 light-years in the best case — and we would need a very sensitive receiver. Besides, at such great distances, the time needed for transmission would eliminate all possibility of interesting dialogue.

Another possibility is to point the receiver in the opposite direction from an exceptional object — a supernova, a pulsar, or the center of the Galaxy. If there is a civilization in that direction, it is probably interested in this object from an astrophysical point of view and therefore it probably observes it. The probability of our being seen is greater. If we transmit in a one-square-degree cone around this direction, the probability of reaching a civilization in this cone (as opposed to the rest of the sky) is 1 in 41,253. In other words, if there are about 50,000 civilizations in the Galaxy, then one of them would hear us if it listened.

If any extraterrestrial civilizations are trying to communicate by transmitting only in those directions where they have detected other civilizations, it is quite possible that we have not been noticed. We have been transmitting radio waves for our own needs for only about 90 years. These waves escape from our atmosphere and travel through interstellar space at the speed of light. They could therefore not have traveled more than 90 light-years — a very short distance, beyond which Earth is a silent planet and presumably empty. In this case, our only possibility of being detected would be to accidentally find ourselves between two evolved civilizations that already communicate.

The best solution seems to be a mixed one: survey the sky systematically with average sensitivity, and aim at a number of well-chosen targets with high sensitivity. This is the strategy adopted in a recent American project which I will discuss below.

Another problem is the likely broadcast time. We have no way of knowing at what time artificial signal transmissions are most likely. In our daily experience, if we want to hear the news on the radio or estimate the time of a public meeting, a course, or a film, we suppose that it is on the hour or the half-hour without much risk of being wrong. It is the same in the Galaxy. Certain "galactic clocks" — novas, variable stars, binary stars — are the sites of natural periodic events on scales varying from several hours to several years. To increase our chances of being detected, we should synchronize the broadcasting of artificial signals with such periodic events.

Such synchronization is difficult, because we have to define the simultaneity in time of two events (e.g., emission of a signal and eruption of a nova) which do not happen in the same place. Indeed, Einstein's theory of relativity tells us that two events that are simultaneous for one observer will not be so for another, since the speed of light is not infinite. The same event will be perceived by different observers at dates which depend on their distance from it. Although the transmission of a signal and the eruption of a nova will be simultaneous for the transmitting society, we will

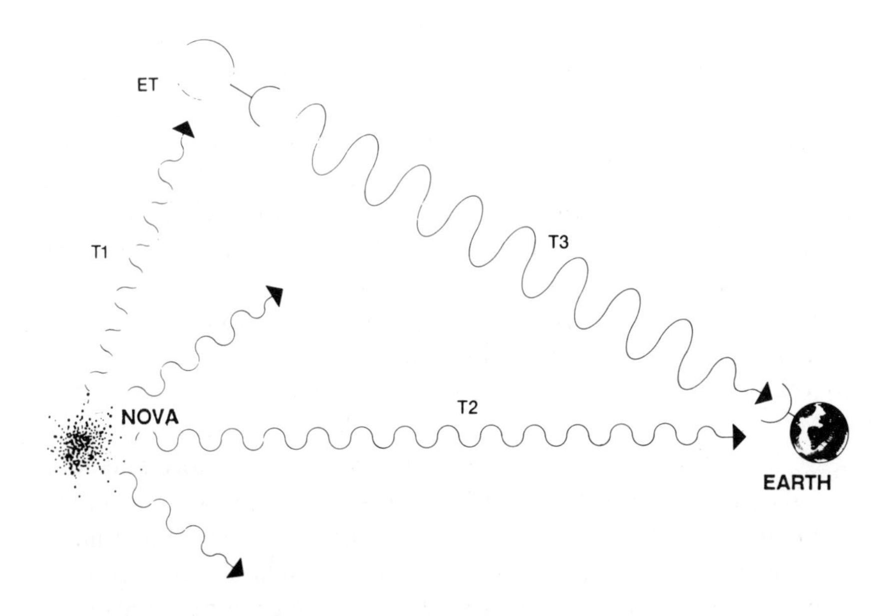

Emission synchronization. The eruption of a nova allows us to set the dates for emitting and receiving messages. The extraterrestrial emitting society (ET) perceives the nova eruption at time t_1; at that moment it sends a message. Earth sees the nova eruption at t_2 and ET's signal at t_3. t_2 and t_3 are not equal because t_1 is not zero.

see them at different dates. The time lag will depend on the form of the triangle formed by Earth, the transmitting society, and the nova.

Novas are binary stars having sudden bursts of activity which make them much more brilliant. In our neighborhood there is, on the average, one every 10 years. The nova in Cygnus (August 29, 1975) was the last one discovered. Intelligent societies neighboring this nova could have found this a convenient moment to transmit signals. The nearer the transmitting civilization is to a line between the nova and the Earth, the closer the eruption of the nova and the reception of the artificial signal will be for us. The idea of synchronizing time by nova is common enough to have been suggested independently by two astronomers — one American, the other Soviet.

The luminosity of variable stars varies according to a cycle which can last from several hours to several years, depending on the star. Detection of artificial signals coming from one of these stars is most likely at the moment when it is least luminous or when it is most brilliant.

Binary stars revolve around each other. The period of this revolution varies from several days to several years, depending on the mean distance between them. The moments when they are closest or furthest apart stand out in their history and could serve to synchronize transmission and reception.

A message transmitted with the purpose of establishing contact with other civilizations should be easy to detect and have a simple code and easily decipherable content. The simplest code is that of Morse signals, invented at the end of the nineteenth century to serve for telegraphic communications: a series of short and long signals. A first message could be a series of numbers, such as the prime numbers (1, 2, 3, 5, 7, 11, 13,...), or else an irrational number (such as π or $\sqrt{2}$) repeated indefinitely. These numbers have semantic value insofar as the signal cannot be generated naturally and requires intelligent intervention.

Another elementary message could take the shape of a drawing. One would just have to draw on successive lines a sequence of signals and silences. The end of each line would be identified by

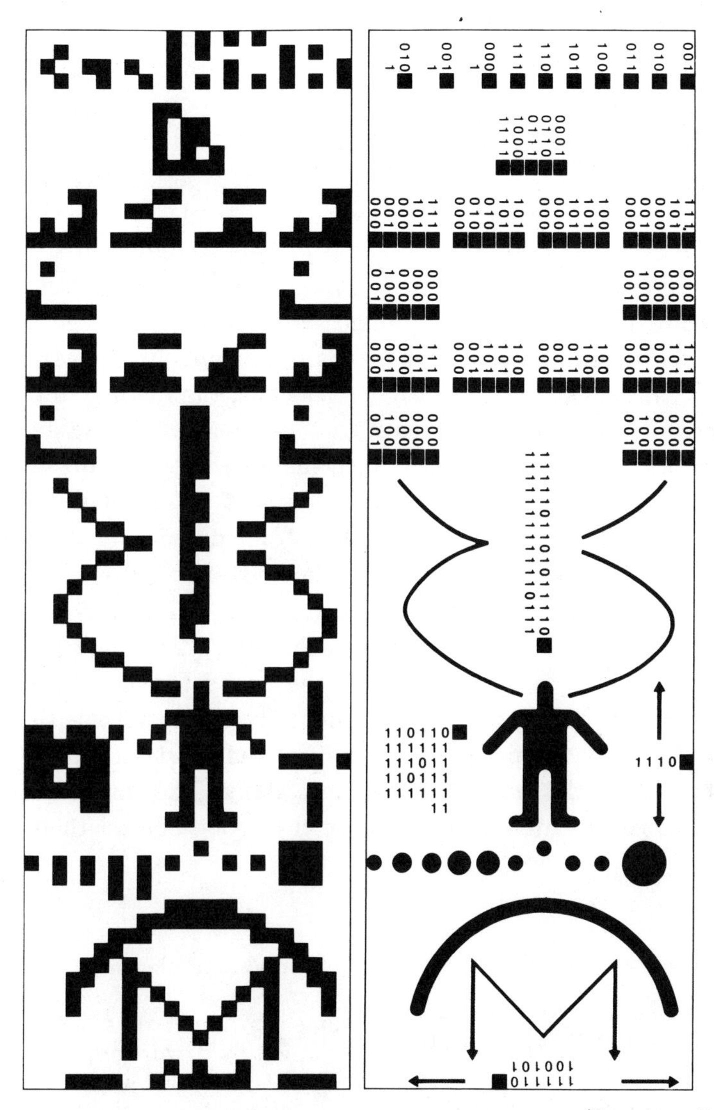

(Encyclopaedia Universalis)

The Arecibo message. The picture on the left shows the message in binary code; the white squares represent zeros, the black squares ones. The message is deciphered on the right. The first line shows the numbers from one to ten; the second, five atoms; the following show the formula of the composition of DNA and its double helix; the last three lines give the global population and depict man and his size, the solar system, and the Arecibo telescope and its diameter.

a special sign — a longer signal, for example. This is the strategy of the message sent on November 16, 1974 by the radio telescope at Arecibo, Puerto Rico, toward the globular cluster M 13. It contained 1,679 characters in the binary system, and it gave in order the numbers from 1 to 10 and the atomic weights of some basic atoms. It then described twelve simple molecules, DNA, the human form, the number of human beings (4 billion) and their size (14 units of 13 centimeters), the solar system (emphasizing Earth), and the radio telescope and its dimensions. Such a message is not very easily decipherable, but perhaps it was meant for *really* intelligent extraterrestrial civilizations. Anyway, the main goal of this experiment was not so much to start a program of interstellar communication as to show that we have the technology to do so.

A first message was sent into space on March 3, 1972, when the American probe Pioneer 10 was launched toward Jupiter. After observing Jupiter, the probe left the solar system at the speed of 11 km/sec in the direction of the constellation Orion, carrying a message engraved on an aluminum sheet 15 by 22.5 centimeters. This message was also placed on the probe Pioneer 11; it is a drawing representing man and woman, the solar system, the hydrogen atom, and the periods and positions of fourteen pulsars with respect to Earth, to locate it in the Galaxy. The size of the human beings is indicated by a sketch of the probe behind them.

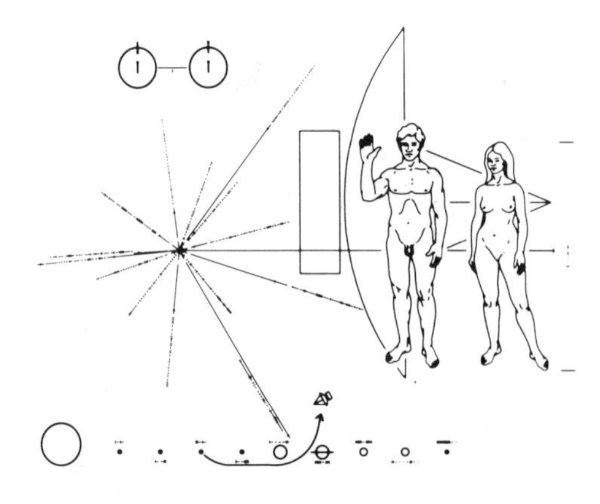

(Encyclopaedia Universalis)

The message placed on Pioneer 10 and 11. The 21-cm line is symbolized at the upper left by a hydrogen atom. The central star shows the directions of the main pulsars seen from Earth, as well as their period of emission. Underneath is a diagram of the solar system. The man and woman are shown in front of the probe.

A drawing is a simple way of communicating. A more sophisticated message requires building a language, a code of communication. The message must then contain a sort of Rosetta Stone to allow possible auditors to decipher it. In 1960, Hans Freudenthal conceived such a language, based on mathematics and symbolic logic, which he called Lincos. Physics and astronomy are sufficiently universal themes to be tackled at the first stages of communication. That they are universal is shown by the experience of the different civilizations on Earth. These ideas are of course very anthropomorphic and assume for our possible audience a gift of abstracting which it may not have. A more realistic attitude is to wait for contact and then to establish a common language by apprenticeship.

If we ever detect signals whose nature is evidently artificial, the task of making sense out of them may prove impossible. This does not mean that interstellar communication is impossible. The mere receipt of a message, even an incomprehensible one, would be a first step toward contact, and would signify the end of our isolation. One can also hope that current research on artificial intelligence will be useful in this endeavor.

11

The Search for Signals of Artificial Origin

The eternal silence of these infinite spaces frightens me.
—Blaise Pascal, *Pensées, III*

The search for signs of extraterrestrial life is certainly the best test of present hypotheses about life in the universe. Taking an inventory of habitable sites, speculating about the characteristics of hypothetical extraterrestrial civilizations, and defining research strategies are indispensable preliminary steps toward successful exploration.

But putting the strategy of the preceding chapter into practice is not easy, because present radio telescopes were not devised for this type of research. First of all, how powerful must an extraterrestrial transmitter be for us to be able to detect its message in the cosmic noise? If it is 1,000 light-years away from us, its power must be between 1 and 20 billion kilowatts. This corresponds to our most powerful radio transmitters. We therefore have a reasonable chance, with our present technology, of discovering an artificial signal if all the other conditions are met — that is, if we observe at the right frequency, at the right moment, and in the right direction. By reversing the above data, we can describe our own transmission possibilities. Our best transmitters can reach about 300 to 1,000 light-years away. There are about 100,000 stars like the Sun in a sphere with a radius of 300 light-years.

Table 11
Ongoing and past searches for artificial radio signals of extraterrestrial origin.

	Researchers (NAME OF PROJECT)	Site	Wavelength (cm)	Target	Duration (hours)
1960	Drake (OZMA)	USA	21	2 stars	400
1963	Kardashev	USSR	32	quasar	
1966	Kellerman	Australia	6–85	1 galaxy	
1968–69	Troitskij	USSR	21; 32	12 stars	11
1968–	Troitskij	USSR	21; 50; 100	all sky	
1970–	Troitskij, Bondar, Starodubtsev	USSR	16; 32; 50	all sky pulsed signal	700...
1970–72	Slysh	France	18	10 stars	
1971–72	Verschuur (OZPA)	USA	21	9 stars	13
1972	Kardashev	USSR	60; 745	all sky	
1972–76	Palmer (OZMA II)	USA	21	674 stars	500
1972–	Kardashev, Gindilis	USSR	18; 223	all sky pulsed signal	
1973–	Dixon, Ehman, Raub	USA	21	all sky	
1974–76	Bridle, Feldman	Canada	1; 3	70 stars	140
1974–	Shvartsman (MANIA)	USSR	5; 4	21 strange obj	
1975–76	Drake, Sagan	USA	13; 16; 21	4 galaxies	100
1975–79	Israel, de Ruiter	Netherl.	21	50 stars (1)	400
1976	Clark, Black, Cuzzi	USA	3.5	4 stars	7
1976–	Bowyer (SERENDIP)	USA	18; 21	all sky	....
1977	Black, Clark, Cuzzi	USA	18	200 stars	100
1977	Drake, Stull	USA	18	6 stars	10
1977–	Wielebinski	FRG	21	3 stars	2
1978	Horowitz	USA	21	185 stars	80
1978	Cohen, Malkan,	USA	1.3; 18	25 globular	60
	Dickey	Australia	18	clusters	20
1978	Knowles, Sullivan	USA	60; 230	2 stars	5
1979	Cole, Ekers	Australia	6	F, G, K stars	50
1979–	SERENDIP II	USA	radio	(2)	400
1979	Tarter, Clark	USA	18; 21	200 stars	10
1980	Lord, O'Dea	USA	0.34	rot axis Gal.	50
1981	Israel, Tarter	Netherl.	21	85 stars	600
1981–88	Biraud, Tarter	France	18; 21	347 stars	700
1981	Shostak (SIGNAL)	Netherl.	21	Gal. center	4
1982	Horowitz, Teague	USA	10.5; 21	150 stars	50
1982	Vallée	Canada	2.8	Gal. center	72
1983–	Horowitz (SENTINEL)	USA	10; 18; 21	all sky	
1983	Damashek	USA	76; 5	pulsars	700
1983	Valdes, Freitas	USA	18	92 stars	100
1983–	Gulkis	Australia	3.7; 12.5	south sky	800...
1983	Gray	USA	21	all sky	
1983–	Cullers (AMSET)I	USA	21		
1983–	Stephens	Canada	19; 21	all sky	
1984	Troitskij	USSR	21	all sky	

(1) Analysis of radio maps of 250 stellar fields to find coincidences between residual radio emission and bright stars.
(2) Analysis of residual radio signals registered during tracking of NASA satellites.

Table 12
Searches undertaken at other wavelengths to detect artificial signals of extraterrestrial origin.

	Researchers	Site	Range	Target	Duration (hours)
1974	Wishnia	satellite	ultraviolet	3 stars	
1979	Freitas	US	visible	(1)	30
1980	Witterborn	US	infrared	20 stars	50
1981–82	Valdes, Freitas	US	visible	(2)	70

(1) Search on 90 photographic plates of objects possibly resembling space probes at the equilateral equilibrium points of the Earth-Moon system.
(2) 137 plates scanned in an attempt to detect objects a few meters in diameter on 23 stable orbits near the Lagrangian points.

Radio astronomers have been listening for about 30 years, with no positive results. Some are already worried about this great silence, but such feelings are premature, because of the insufficient numbers of observed sources, explored frequencies, and hours spent listening. New receivers conceived especially for such research are being built in the United States and will no doubt give a more categorical answer in the years to come.

Table 11 describes astronomical observations already completed or in progress for detecting artificial signals. Most were done in the radio domain, but with conventional receivers, which rarely permit a complete scanning of all frequencies. The number of hours spent listening is not very significant, since some recent observations are much more efficient than older ones thanks to multichannel analyzers. For example, the information processed by Drake in 200 hours of observing time in 1960 could now be handled in less than a second by the 1,008-channel receiver in Arecibo.

Sometimes these observations are passive, that is, they are by-products of conventional astrophysical or astronautical observations. For example, the main purpose of the observations of Cohen and his collaborators was to detect oxygen compounds in globular clusters; the detection of possible signals was secondary.

It was during classic radio-astronomical research that Jean Heidmann of the Paris Observatory noticed a weak stellar object

(Sydney Harris, ed. Belin (Paris))

"The message from space asks us to stop sending incomprehensible messages into space."

at the position of a wide-band radio source in the neighborhood of a Markarian galaxy.[1] Subsequent spectroscopic observations showed that it was a G-type star, like the Sun. If this star, and not a quasar behind it, is indeed responsible for the radio emission, it must be a million times stronger than the "calm" Sun (outside periods of radio bursts). According to Heidmann, this intense radiation could be explained by a gigantic parabolic reflector concentrating the star's natural radio emission into a narrow beam, like a beacon. But more accurate radio observations in 1986 showed that the source is not exactly at the star's position.

The Nançay decimetric telescope, in Sologne, France, is well adapted to the search for artificial signals. With a collecting surface of 200 × 35 meters and a receiver tuned to the wavelengths of 18 and 21 centimeters, it can detect a 10-million-megawatt monochromatic signal emitted 40 light-years away from Earth in 1 second. The receiver's 1,024 channels let us simultaneously analyze the same number of 50-hertz-wide frequency bands. To detect a weaker signal, we just have to listen (or "integrate," in radio astronomers' jargon) longer.

François Biraud, a radio astronomer at the Meudon Observatory, made six series of observations in collaboration with Jill Tarter of the NASA-Ames Research Center, listening for a total of 700 hours between 1981 and 1988. They were looking for artificial signals coming from stars that could not be observed at Arecibo.[2]

The targets of this particular search were 347 stars of types F,G, and K located less than 75 light-years away. In June 1981, they observed 102 stars. Each star was observed for 25 seconds at four wavelengths close to 18 centimeters. The star RGO 216, which gave suspicious results, was observed once again in April 1982, but with negative results. Technical problems prevented the second set of observations, made in October 1982, from being reduced. In August 1984, 155 stars were observed at the wavelengths of 18 and 21 centimeters, with no positive results except for one false alarm

1. Markarian galaxies are characterized by an excess of ultraviolet light.

2. Observation at Arecibo is limited to declinations between − 39° and − 2°. At Nançay, one can observe from the north celestial pole to a declination of about − 40°.

caused by an emitter on Earth. In April 1987, star RGO 243 gave a strong signal, which again turned out to be of terrestrial origin.

(L. Gouguenheim)

The large antenna of the Nançay radio telescope. The collecting surface is formed by a tightly woven screen, 200 meters long and 35 meters wide, that can be moved in declination. To compensate for the Earth's rotation, the telescope's focus moves during observations of radio sources.

In all, 344 stars were observed, among which 95 were actually multiple stars. No signal with the expected properties of an extraterrestrial signal was detected. These observations nevertheless had value — they identified inadequate observing procedures and technical problems whose solution will improve future SETI observations and their analysis.

In all the preceeding searches, the artificial signal was assumed to be monochromatic. But Biraud believes that wide-band signals hold real possibilities. A monochromatic signal oscillates regularly in amplitude with time; its form is sinusoidal. However, with a wide-band signal, the time modulation can be any way at all. It can also be altered by its passage through the interstellar medium.

(NASA)

The Cyclops project. This project from the 1970s, now abandoned, planned 100 antennae, spread over an area 8 kilometers in radius. At the time, a total collecting surface of 7 to 20 square kilometers seemed necessary to collect artificial signals in the cosmic noise.

(NASA)

Detail of the Cyclops project. Each antenna has a diameter of 100 meters. On the right is the control and signal-processing building.

If the form of the wave is not known beforehand, the computation time needed to detect the wide-band signal becomes very long.

The radio telescope of Ohio State University is of the same type as that at Nançay. Since 1969, it has been used only to search for artificial signals — which explains its nickname, "the big ear." In August 1977 a remarkable signal was registered at the telescope, the most intense alarm in 8 years, but this "Wow!" signal[3] never appeared again on the computer charts, and was never explained. In 1985, a golf club wanted the telescope site. Only a concerted effort by the international astronomical community prevented the telescope from being turned into scrap.

NASA's grandiose Cyclops project of the 1970s has since been abandoned for less ambitious but more realistic programs using existing radio telescopes. NASA's most recent project of artificial-signal research relies on revolutionary new receivers. A spectral analyzer could detect and analyze signals in 8 million frequency bands at a time, in real time (i.e. during observation), each band having a width of 1 to 32 hertz. This kind of receiver is considered necessary because the hypothetical artificial signals would probably be emitted in a narrow band, and we cannot know in advance how narrow it is. Widening the band to have fewer channels to analyze would serve no purpose; if the emission band is too wide, the signal is spread out and buried in the Galaxy's background noise.

NASA's new strategy is, on one hand, to observe 1,000 stars similar to the Sun and less than 80 light-years away at frequencies from 16 to 21 centimeters with a resolution of 1 hertz (in other words, 2 billion channels to be analyzed for each star), and, on the other hand, to sweep the sky systematically in a wider range of frequencies with a smaller resolution, spending only a few seconds at each position in the sky at a given frequency. The preliminary budget for the program (with an analyzer of 74,000 channels) is estimated at $12 million over 5 years (0.04 percent of NASA's budget).

Researchers at the University of California have automated the technique of passive observation. Their program is a parasite in

3. at right ascension 19h17m and declination – 27° 3'.

the sense that it is grafted onto classical astrophysical observations, but without interfering with them. The data obtained by radio during conventional research is analyzed automatically. The choice of frequencies, targets, and observing periods is thus imposed by the principal research program, which concerns objects of astrophysical interest. The spectra obtained during observations are stocked on magnetic tapes, and then a microprocessor "listens" to them to detect signals in a narrow band. The detection threshold is such that about 4,000 events are discovered monthly. Each event is identified by its coordinates (right ascension, declination, hour angle, azimuth) and a date. Any signal that is not a statistical fluctuation of the background noise or a parasite signal of earthly origin might be detected several times. The strategy is to try to recognize systematic behavior of the detected signals, in terms of coordinates. A first series of observations were made over 35 days in 1980 during conventional observations of giant red stars harboring maser sources.[4] Analysis turned up about 4,000 signals that went beyond the average fluctuations of the background noise of the sky and the receiver by a factor of 4.5. The origin of several groups of signals was the receiver itself. Sixty remain unexplained. Three or four are probably due to statistical fluctuations of noise. It would therefore be useful to once again observe in the directions the antenna was pointing during these events.

Amateurs are active in many areas of astronomy, and even in the search for artificial signals. An engineer-programmer in Chicago, Robert Gray, recently built an automated telescope dedicated to this research. It is a parabolic antenna 4 meters in diameter mounted on an azimutal radar frame and moved by a 0.5-horsepower motor. The receiver is set to the wavelength of 21 centimeters and to 256 channels whose bandwidth can vary from 1 to 100 hertz. This enables it to detect a 1,000-megawatt signal emitted 50 light-years away by an antenna 100 meters in diameter, in 16 seconds. However, the telescope's small diameter gives

4. A maser acts like a laser, radiating monochromatic light in a very intense way after being stimulated by a physical process (which can be natural).

it a low resolution, so it cannot localize possible sources very precisely. The telescope is piloted by a microcomputer which makes the observations and reduces the data automatically. It generally works 6 to 12 hours a night. It could also work during the day, but then interference from terrestrial radio emissions is more frequent. Because of the narrow band of the channels, Gray's receiver can detect only the most intense radio sources. These are used as routine objects in observation to calibrate other possible signals. This amateur works mostly in an area of the sky in which the Ohio State University radio telescope detected suspect signals one day. No positive results have yet been obtained, but this probably proves that the instrument does not detect terrestrial parasites.

The extraterrestrial origin of a monochromatic radio signal becomes certain if there is a slight drift in frequency (0.1 hertz per second) of the signal, due to Earth's rotation with respect to the emitter, and if it appears 4 minutes earlier every day.

Under the label AMSETI, a group of radio amateurs in the Silicon Valley of California have tuned a spectral analyzer to a wavelength of 21 centimeters and connected it to a satellite-TV dish.

Another group, the Terrestrial Research Institute in Canada's Northwest Territories, has acquired two ionospheric antennas 15 meters in diameter and constructed a low-noise receiver for frequencies at the "water hole."

The perseverance of these amateur groups will depend on the enthusiasm of their members and especially on their financial resources.

The search for artificial signals is only beginning. But perhaps it is already too late. The use of more and more radio emitters on Earth and even around the Earth is making listening ever more difficult. Some frequency bands which are crucially important for radio astronomy are protected. Among these are the emission frequency of neutral hydrogen (1,420 MHz) and that of the hydroxyl radical (1,660 MHz). These are exactly the two limits of the "water hole," where radio astronomers hope one day to detect artificial radio signals. But starters for gasoline motors and remote garage-door openers are not included in the protection measures. A radio telescope could detect the emission of such a device 40,000 kilometers away. The Soviet Union's Glonass navigation satellites, which transmit in the neighborhood of 1,600 Mhz, prevent the Nançay radio telescope from observing at that frequency about 12 hours a day. When all 24 Glonass satellites are in orbit, around 1992, that frequency will be completely blocked for Nançay. Radio astronomers will then have to establish observatories on the hidden face of the Moon.

12

The Debate within Society

Man's obscure relentlessness to re-create the world is not in vain because nothing becomes presence beyond death, except for re-created forms.
—André Malraux, *The Imaginary Museum*

Up to this point, my account has been organized around elements of scientific knowledge in very different fields. I presented the astronomical, biological, paleontological, and sociological aspects of life on Earth that could be universal in character. I then made an inventory of astronomical knowledge and research that could help convince us that life exists elsewhere in the universe. Before drawing the conclusions and implications of this research for our view of the world, and possibly for our future, I will discuss how the search for life in the universe has been perceived and accepted in the recent past by astronomers and by the general public. I will put this research into its historical context and talk about the debates it has generated among astronomers. The way the general public sees it is revealed by the results of several opinion polls. As we shall see, public support is important in democratic societies in ensuring financial support for this research. The purpose of this analysis of the currents of opinion that have influenced this research, or could do so in the future, is to situate our account in a wider context than the relatively narrow circle of researchers who have contributed to it.

The starting point for modern research on life in the universe was an article published in 1959 in the British journal *Nature* by two scientists at the Massachussetts Institute of Technology.

Giuseppe Cocconi and Philip Morrison suggested that the best wavelength for interstellar communication was 21 centimeters,the wavelength of the radiation of neutral hydrogen, the most common element in the universe. At the time, the article raised little interest and no polemics such as those that usually follow the publication of new and provocative ideas. But Frank Drake, then a young radio astronomer at Green Banks, revived the idea in 1960 for his Ozma project, in which he tried (unsuccessfully) to pick up artificial radio signals from the direction of two of the Sun's closest neighbors. Although Cocconi never continued this work, Drake and Morrison remain enthusiastic about such work, even while they also work on more classical themes.

Ten years went by before other radio astronomers began to listen to stars. The theoretical background was missing, as was a synthesis of scientific knowledge to establish the validity of this new field of research. Indeed, why talk about interstellar communication if there is no one to communicate with?

The first book examining the possibility of finding intelligent life in the universe through a complete inventory of scientific knowledge came from the Soviet Union. The author of this work, published in 1962, was Iosef Shklovskij. An English-language edition, completed by Carl Sagan, was published in the United States in 1966. At about the same time, a French translation of the original book appeared under the title *Univers, Vie, Raison.* This title evokes well the structure of the book: an astronomical introduction, a study of life on Earth, and finally the possibility of intelligence in the universe. In particular Shklovskij suggested that Phobos and Deimos, Mars' two satellites, were hollow, and therefore artificial. We now know that the actual reason for their apparently low masses is their irregular shape.

The search for life in the universe has been recognized as a legitimate interest of the scientific community by the International Astronomical Union (IAU). Commission 51 of the IAU, created in 1981, already has 250 members, 5 percent of the IAU's total membership. Its title is Search for Extraterrestrial Life. Its purposes are to search for planets in other solar systems, to investigate the evolution of planets and their ability to shelter life, to search for biological-type molecules and study their formation, to search for artificial radio signals of extraterrestrial origin, to search for evidence of other advanced civilizations, to conduct a spectroscopic search for primitive life around other stars, and to coordinate international activities on these topics.

The first IAU symposium on this subject, entitled "The Search for Extraterrestrial Life: Recent Progress," was held in Boston in June 1984 and attracted about 150 participants. It was not the first international conference on the subject; others had already been held in the United States and also in the Soviet Union (Byurakan, 1971; Tallin, 1981), but they were not sponsored by the IAU and were the results of individual initiatives.

Despite this official recognition, the Western scientific community is far from being unanimous about the usefulness of searching for intelligent life in the universe. It is normal that scientists disagree about the origin of life on Earth, about the causes of periodic catastrophes and about the reasons for the great silence. It is in fact this difference of opinions and hypotheses that makes this topic so interesting. But, if there is consensus on the usefulness of searching for other planetary systems, or for signs of life on other planets in our solar system, astronomers are divided as to the advisability of looking for intelligent life and artificial signals now. They question the research itself, rather than its possible outcome.

Dissent also appeared in the Soviet Union, where astronomers' important contribution to extraterrestrial-civilization research, at least to its methodology, seemed to give an impression of consensus about the usefulness of this research. In 1977, a debate arose between Shklovskij, who had clearly become pessimistic, and the supporters of this research in Eastern Europe, including Nikolai Kardashev, Vladimir Troitskij, and Polish science-fiction author Stanislaw Lem. It was published in several episodes in the Soviet journal *Znanie-Sila.*

To understand the reasons for this disagreement, it is necessary to know how scientists work. A scientist bases his research on solid knowledge and justified hypotheses. Sometimes he follows his intuition rather than logical reasoning, and this attitude can be fruitful. But the investment required in telescope time and in man-hours of labor at the basis of astronomical research often prevent the exploration of areas with no scientific premise (also called "going fishing"). The search for artificial radio signals is a good example. In this particular case, the choice of experimental parameters is so great, the cosmic haystack is so huge, that the chances for success are very small, especially since we do not even know if "they" are emitting radio signals. Should such an effort be continued, given the investments in telescope time and personnel, if the first results are not promising?

The economic consequences of this debate are obvious for those who know how scientific research is managed. Funds for research and telescope time are allocated only on the basis of their scientific merit; renewed financing depends upon the results obtained. If the scientific interest of a research topic has not been recognized, or if the research has not been fruitful, work on the topic will not be encouraged. This explains why telescopes used for extraterrestrial-life research are usually the property of small groups, and why most observations are shared.

The negative results of all attempts to detect artificial radio signals gradually caused a certain number of scientists, including Shklovskij, to review their arguments in favor of intelligent life in the universe and to moderate their enthusiasm.

This skepticism reached its peak at the beginning of the 1980s when Frank Tipler published an indictment of the research in a series of articles. He even questioned the motivations of the principal proponents of such research. Readers' responses and the simple fact that Tipler's articles were accepted by journal editors show a clear evolution in ideas about this research.

Tipler's reasoning is simple: If the whole Galaxy is inhabited by civilizations slightly more technologically advanced than ours, and who are even minimally motivated to explore or come into contact with other intelligent beings, at least one of them would already have sent its probes into our solar system. Since this is not the case, we are alone and the research is useless.

One needs very strong arguments to be so categorical. Tipler claims that it is possible to explore the Galaxy in less than 300 million years with our present technological capabilities in space probes and with computer capabilities which will be available within the next few centuries. Such futuristic computer technology is needed to produce von Neumann probes — robots capable of repairing themselves and reproducing from materials gleaned in planetary systems. Furthermore, Tipler asserts that a civilization which is not motivated to explore the universe does not meet the definition of an intelligent civilization.

Tipler's argument is based on predicted developments that may not take place. And even if the absence of probes in our space environment excludes the existence of more advanced civilizations than our own, what about those at our level who are trying to make radio contact? In my opinion, it is not possible to be as categorical as Tipler without having completely explored the universe looking for artifacts and artificial signals. At the present rate, several decades are probably necessary. Philip Morrison considers that we shouldn't abandon such research without one or two centuries of efforts.

Tipler is nonetheless right when he questions the unscientific motivations of his colleagues. Quoting from the writings of Carl Sagan, Fred Hoyle, and Frank Drake, he shows that they attribute to extraterrestrial civilizations a Messianic role which will save us from coming technological disasters. Drake even hopes that advanced civilizations will give the Galaxy's younger civilizations the secret of their immortality. I should also mention Michael Papagiannis, who ends one of his articles on the search for extraterrestrial civilizations with an irrefutable proof of the existence of a creator of the universe. Tipler also had to defend his first article against Sagan, who, as a referee, refused its publication in two journals successively. The contents of Sagan's reports made Tipler feel as if he were involved in a theological debate. Tipler relates that Michael Hart, author of a study on continuously habitable zones, had the same trouble publishing his articles, especially in the United States.

Even though Tipler's arguments are debatable, they at least had the merit of forcing the principal authors in the field to reexamine their motivations and arguments and to find solid justification for this research.

The publication of Tipler's articles had a catastrophic effect on the financing of artificial-signal research. After becoming acquainted with one of these articles, Senator William Proxmire managed to persuade the US Senate to do away with NASA's modest subsidy (a few million dollars) in 1982. The effect of this negative publicity was immediately named the "Proxmire effect" by researchers with a sense of humor. The Proxmire effect is one of the reasons — difficult to integrate into research procedure — why we haven't yet intercepted any artificial signals of extraterrestrial origin: the decision-making bodies of other civilizations may have their own Proxmires.

An energetic response to this pessimism was obviously needed. Carl Sagan took the initiative by drafting "Extraterrestrial Intelligence: An International Petition" and getting it signed by 69 researchers from the international scientific community. It was published as a letter to the editor of the American journal *Science* in 1982. Its purpose was to win over the greatest possible number of scientists, preferably eminent ones, to the idea that the time had come to answer questions about intelligent life in the universe with modern technical means, and to show other scientists, Senator Proxmire, and the public that some consensus existed on this matter.

In the petition, after a few preliminary remarks on the real possibility we have of communicating by radio signals with other civilizations in the universe if they exist, on the modest cost of such a research program (a few million dollars a year), and on its urgency (interference by terrestrial radio emissions can only get worse), Sagan attacks the principal obstacle to the advisability of this research: the absence of visible signs of "their" presence in the Galaxy, and particularly in our solar system. According to him, absence of proof is not proof of absence, and such a research program has the merit of not attributing to possible extraterrestrials competence that we ourselves do not possess. Here Sagan probably refers to the hypothetical von Neumann probes that are at the basis of Tipler's pessimistic arguments.

Sagan then draws our attention to the signers, who are from very diverse scientific and geographical areas. Besides those researchers directly or indirectly involved in the research that I have already mentioned (Drake, Morrison, Tarter, Shklovskij, Raup, et al.), we find the names of several Nobel laureates (Chandrasekhar, Crick, Herzberg, Pauling), managers of science in the United States (DuBridge, Field), a French radio astronomer (Jean Heidmann), and some British scientists (including the physicist Stephen Hawking). Even if they are divided on the probable number of other civilizations, on the chance of success of a radio search, or even on the arguments in favor of "their" existence, all the signers are convinced that the only valid test of this hypothesis is experiment.

This petition could not leave Tipler indifferent. In a letter to the editor of *Science*, he criticized it on two points: If extraterrestrial civilizations are on the same scientific and technological level we are on (and thus not systematically transmitting radio signals for interstellar communication), our chances of detecting them are practically nil. If the results are negative, the impact of this research will be nil, since we would not have proved their absence — perhaps they abandoned radio communication in favor of a more efficient method.

But, Sagan answered, we have already sent a message (toward the globular cluster M 13). And if we choose not to transmit, we at least have the technical means of doing so, if not the financial ones. Jill Tarter's response is more instructive: the weakest signal that we can now pick up must come from a 1 to 20 megawatt emitter, if it is located 1,000 light-years from Earth. This power is comparable to that of the radio telescope at Arecibo. There is thus no need to be as pessimistic as Tipler, who admits in another article that a radio search would be a test of his pessimistic views.

It is all right to look for microscopic life on the surface of Mars, or to try to pick up artificial radio signals coming from outer space, but discovering bacteria in the interstellar medium is considered sacrilegious. This is the conviction that Fred Hoyle and his colleagues at Cardiff University have acquired in recent years.

Hoyle is one of the most brilliant British astrophysicists. He is one of the founders of the stationary-universe theory, and his name is linked with some of the great astrophysical discoveries of the 1950s. For many years, he and his colleague Wickramasinghe have been interested in interstellar dust and in the absorption of starlight as it passes through clouds of such dust. They first tried to attribute the extinction to carbon particles, then to solid hydrogen, then to magnesium oxides, but all without satisfactorily reconciling laboratory experiments with astronomical observations. Only after a long search in the chemistry literature did they finally, in 1977, get the idea that interstellar grains could be organic products, such as polysaccharides. Using this hypothesis, there was remarkable agreement between laboratory and telescope data, and their problems of having this new idea accepted began. The difficulties did not arise so much from the idea of organic interstellar dust as from the accompanying speculation about life in the universe and panspermia. Since then, their publications about panspermia have been distributed as preprints and low-circulation brochures, outside the traditional network of scientific information — scientific periodicals. The latter refuse to publish what they consider, rightly or wrongly, to be conjectures without solid scientific bases.

I mentioned in chapter 5 the denials that were made concerning interstellar molecular clouds. Hoyle responded by directly attacking the authors: "If Davies and collaborators had answered this question with an open mind instead of looking for mistakes at any price...." This polemical tone, unusual in scientific literature, shows either that Hoyle had lost patience with his critics or that his convictions had made him lose the objectivity and detachment necessary for scientific research. This last hypothesis seems to be correct, because, instead of appearing in *Nature*, in compliance with the right of response, his answer appeared in *Astrophysics and Space Science*, whose editorial policy is very liberal. As a commentary, the editor of the journal followed this response with a letter from two Japanese researchers who had tried to reproduce Hoyle's laboratory experiments. Despite all their efforts, they were unable to find the absorption properties Hoyle claimed to have detected in the ultraviolet range.

A second debate, this time in the pages of *Nature* during the summer of 1986, arose when Hoyle and N.C. Wickramasinghe criticized the absence of references to their work in an article by J.

Mayo Greenberg on Halley's Comet. They attributed this absence to bias and unfairness on Greenberg's part. They also accused *Nature* of having suppressed a reference to one of their works in an article of which they were the authors. According to them, the reason for this suppression was to repress all criticism of Darwinism. In his response, John Maddox, one of *Nature*'s editors, noted that the work in question referred to their panspermia theory and that it was precisely their proselytizing for this still purely speculative theory that was putting scientists off their other very valid work, such as the deduction that comets contain organic molecules. Their vision of microscopic life distributed throughout the universe by dust may turn out to be correct, but for the moment it is only conjecture.

All these heated debates show that scientists do not always silence their own feelings and convictions in this research, which obviously has implications beyond the scientific domain.

And how does the public perceive the search for life in the universe? In an April 1966 Gallup poll in the United States, the following question, among others, was asked of 1,500 people: "Do you believe that there are people a bit like us living on other planets in the universe?" Positive replies were given by 34 percent, 46 percent said No, and the rest had no opinion.

In November 1973, in another Gallup poll, the rate of positive answers was 46 percent, versus 38 percent negative. The conviction that there is extraterrestrial life was more prevalent among more educated and wealthier people. 58 percent of those who answered Yes had been to college, versus 29 percent who had never been to high school. Men were more affirmative than women (48 percent vs. 45 percent), Catholics more than Protestants (46 percent vs. 43 percent). The difference of opinion was greatest in the different age groups — 66 percent of those under 25 answered yes, against 30 percent of those over 50. Since young people will form tomorrow's society, this favorable opinion might be decisive for the search for life in the universe.

In 1981 Williams Bainbridge, a sociologist at Harvard University, questioned 1,465 students at the University of Washington, where the future elite of a whole region is formed. The ideas of these young people might influence those of a large segment of the American public, which in turn will influence the financing of future space exploration, and most particularly the search for life in the universe. To the question "Should we try to communicate with intelligent beings on other planets?" half of the men said Yes, versus 35 percent of the women (average: 42 percent). The statement "Intelligent life probably exists nowhere but on our planet" was massively rejected; only 8 percent of the men thought so, as did 12 percent of the women.

The answers to a series of questions about space exploration suggest that in the minds of young people the search for extraterrestrial civilizations is associated with the distant purpose of space colonization, rather than with the more immediate economic and military results of space exploration, such as meteorological and communications satellites and the Strategic Defense Initiative. This is an important point. If the excessive militarization of space might make public opinion hostile to any exploiting of interplanetary space, it will probably have no effect on public support for the search for intelligent life in the universe.

Another aspect of the Bainbridge study revealed that positive attitudes toward the search for intelligent life are more linked to sophisticated technology than to war and national defense. Those who distrusted the search for life in the universe were also those who distrusted advanced technology. In the future, it will be important not to emphasize new technology as promising possible progress in the search for intelligent life, but rather to stress the social and cultural benefits of this progress.

According to this study, interest in the search for life in the universe is linked to curiosity about the vast universe around us. Astronomical progress might thus encourage popular support for our research. On the other hand, students interested in communications, foreign languages, and sociology did not seem to want to extend their interest past Earth. Among students of the social sciences, only anthropology students were ready to be interested in extraterrestrial civilizations.

The search for life in the universe should bring answers to some existential questions, entering a realm usually occupied by religion. It is therefore not surprising that Bainbridge's study revealed religious opposition to this research, especially among Protestants. Only 38 percent of them thought we should try to communicate with extraterrestrial societies, versus 42 percent of Jews, 44 percent of Catholics, and 50 percent of students with no religion. (Remember, 42 percent of all students approved of this research.) The analysis of questions about religious practices showed that very religious students were less likely than the others to approve of space exploration, and especially the search for intelligent life. The especially strong opposition among Protestants comes from the fact that our research questions the literal interpretation of the Bible. Another explanation would be the traditional xenophobia of the most sectarian Protestants.

The popularity of the topic of life in the universe can only benefit from the results of space exploration, without really suffering from its negative aspects. Enthusiastic support among the young is a decisive factor in favor of future financing of this research.

13

Man's Place in the Universe

What man thinks of himself, that is what decides or rather indicates his destiny.

—Henry David Thoreau, *Walden*

The search for life in the universe is just beginning, and I can only end this essay with a question that now appears as one of the purposes of this research: What is our place in the universe?

This research has many objectives. It has revealed to us the richness, the variety, and the possibilities of the surrounding universe, and it will no doubt lead us to discover new marvels in the near future. This research also benefits other scientific fields. It is a multidisciplinary science where scientists of all fields share their knowledge to look for the answer to a question whose implications reach well beyond the realm of science. I have emphasized the contributions of biologists, paleontologists, sociologists, and astronomers in all subfields, from radio astronomy to the study of planetary atmospheres and stellar kinematics. Such a collaboration will no doubt open new horizons to the researchers in all scientific fields involved.

If all these scientists do not provide the answer to the question of life in the universe, they will nevertheless have contributed to our knowledge of life on Earth and of our place in the universe. The attempts to generalize the concepts of intelligent civilization and language will put our destiny in a new perspective, as well as the concepts of life and intelligence. This research will also favor the birth of new sciences, such as bioastronomy, the science of life in the universe.

The astrophysical by-products and applications of a program of this kind will certainly be numerous. Systematic surveys of the sky in narrow frequency bands have never been made. It may well be that astronomers listening for artificial signals will discover new sources of monochromatic radiation in stars and the interstellar medium, transient radio sources, etc. Isolated clouds of neutral hydrogen have already been discovered.

There is also technological fallout. The search for monochromatic radio signals requires the technical development of ultrasensitive multichannel or scanner-type receivers and the treatment of the intercepted signals in real time. These new techniques could be used in various fields of classical astrophysics such as radio spectroscopy, the detailed study of pulsars, and measurement of the internal velocities of galaxies, the study of supernova remnants, and so on. They could also be applied to communication with probes, which we will certainly send further and further away from Earth.

As our very modest physical place in the universe progressively becomes evident, we are led to wonder whether life could possibly be unique. Indeed, if we are the sole living beings, the only inhabitants of this universe, why does it have to be so large? This question is poorly stated, because it attributes a goal to the universe — that of being a host to life. For scientists, the universe has no purpose. (More precisely: the question of purpose is not relevant to the field of science.) However, the appearance of life is tightly linked to observed properties of the universe. If the universe were slightly different, it is likely that we would not exist. Any minor modification of a parameter governing the structure of the universe has a deep effect on its evolution, and hence on that of life, be it terrestrial or universal.

The universe has been expanding for about 15 billion years, and this expansion is slowed down by the force of attraction of the matter it contains. One may very well conceive of a universe where all the fundamental constants of physics are different, a universe that would thus have evolved in a totally different manner. But any other universe cannot give birth to planets and to life as we know it.

If the constant of gravitation were larger, stars would not be bigger than the Earth and would burn up in less than a year. If it were much larger, the universe would stop expanding and start contracting before stars had time to form. In the opposite case of a low value of the constant of gravitation, the expansion would proceed too fast for the forces of gravity to keep stars together and induce nuclear burning in their cores.

The variety of observed stars, convective as well as radiative (that is, stars from which heat is carried outward by convection or radiation), is possible only for a very specific choice of these constants. For any other choice, the stars would be radiative and planets would not be able to form around them, or else the stars would be very convective and the heavy elements too well mixed in their interiors to produce supernova explosions, which redistribute these elements in the interstellar medium.

The intensity of nuclear forces governs the chemistry of matter and of the living world. If these forces were weaker, only hydrogen would be stable; there would be no more complex compounds, thus no Earth, and no life.

In short: the immensity of the universe is a necessary condition for the formation of stars like the Sun, a prerequisite to the appearance of life. The Sun and its string of planets could not possibly constitute a universe by themselves. They could not have formed without the billions of other stars of our Galaxy, or without the billions of other galaxies that populate our universe. This kind of argument has lead to the *anthropic principle*, which states that the presence of life, and of man in particular, has an explanatory value for the history of the universe.

The choice of the constants of physics would be dictated by the necessity of an observer, without whom the universe could not exist and would only be a probability (in the sense of quantum mechanics). This is, in my opinion, more a philosophical than an astrophysical problem: Can reality exist without a consciousness that is aware of it? It would be much preferable to reach a unified theory of physics in which the observed values of the constants of physics have a rational explanation, so that a universe with different values of these constants would simply be inconceivable.

In the most recent inflationary theory of the universe, produced by the Soviet cosmologist Andrei Linde, the universe is infinite and composed of regions separated by uncrossable horizons. The constants of physics have specific values in each isolated region of the universe. Complex evolution of matter takes place only in the regions where the constants have the right values. In that case, it may be only a coincidence that our universe has the properties that we know, which allow us to exist.

This kind of reasoning is, in my opinion, the most fruitful aspect of the search for life in the universe. Indeed, whatever the outcome, I expect that the search will bring a better understanding of the role of life on Earth. An inventory of the vital resources of the universe places our existence in a wider perspective, and this interrogation of the universe is a dialogue during which the universe sends us back a picture of ourselves. This is for me the main benefit of the search. It gives us a better understanding of our place in the universe.

Now let us examine the practical consequences of a concrete result from our search. If we are not alone in the universe and our detection has positive results, one immediate consequence will be a lesson in modesty: not only is the Earth not the center of the universe, but there is nothing special about the civilization which inhabits it. In the long run, having established contact, we would have a new perspective on our form of life, its origins, and its future. We would learn which social forms and structures are most apt to survive. New aesthetic accomplishments would enrich us. We would become full members of the Galactic Club and would share in the cultural, historical, and scientific heritage accumulated over billions of years by various civilizations.

More pessimistic consequences of contact could be the invasion and exploitation of Earth. Fortunately, the enormous interstellar distances probably protect us from such a catastrophe. Subversion is a more subtle danger. Under cover of teaching or technico-scientific assistance, "they" could make us produce the means for them to take over the Earth. In this case, the only defense is a good measure of skepticism, which most scientists do not lack.

There is also the possibility that the simple contact with a highly superior society would give us "culture shock." Whatever the intentions of an intelligent civilization toward us, our own civilization will not be the same after contact. In particular, if "they" are of a higher technical and cultural level, we might risk losing our identity, just like the so-called primitive societies which are being integrated into Western civilization. However, without physical contact such effects are less likely to occur. Furthermore, for cultural interaction to be effective, the two cultures should be reasonably similar. Without wishing to minimize the risks of contact, I think they are minimal at the moment of detection; they increase as quantitative information is exchanged.

Among the scenarios of contact, some probably belong to the realm of science fiction, a literature which owes its success to fascination with life in outer space. Some promoters of bioastronomy live in the religious expectation that such contact would be the salvation for our civilization. Such an attitude is completely foreign to me. I am convinced that, if one day we should discover some form of extraterrestrial life, the surprise will be total. It is therefore rather useless to try to imagine this contact any further. The pictures of the solar system sent back by the Voyager probes have already shown us a world of unexpected variety, and I have no doubts that the universe will surprise us yet again.

Beyond all the benefits that I have suggested, there is a much deeper motivation for wishing to learn one day that there are indeed living beings elsewhere in the universe. An inhabited universe immediately takes on a broader significance to our eyes. All recent progress in astronomy has revealed to us a universe of unfathomable dimensions, criss-crossed by all kinds of lethal radiations and particles, populated with diabolical objects such as quasars, black holes, and pulsars — in short, an inhuman and thus unintelligible world. This image would weaken considerably if we knew that somewhere in our Galaxy there is at least a planet with a few flowers! Even more, an intelligent presence would provide an initial answer to our existential questions.

A negative result would be just as valuable if we were able to turn it to our advantage. If all the efforts of scientists to find life elsewhere should remain fruitless, future generations will finally be convinced that they are alone. Once they get over the dizziness of being confronted by this infinite emptiness, they would have to meditate on the meaning of their unique presence, and on their role as solitary witnesses of a silent universe. But have we always been alone, and will we always be so, or are we the first samples of a life that is emerging in the Galaxy? What does this little planet, this little island of life lost in the immensity of the universe, mean? This is the starting point of a reflection I invite you to pursue.

Appendix: The Galaxy

The heavens seem to be uniformly studded with stars; however, if we examine the sky more carefully, we can see a lighter, wide band which crosses it, and with a telescope we can make out an infinite number of stars: the Milky Way. Astronomers call it the Galaxy, with a capital letter to distinguish it from other galaxies. Since I often mention the Galaxy in the pages of this book, I will give a more detailed description here.

The Galaxy is a gigantic disk, composed of stars and of gas, which spins in an orderly way around its center, the nucleus. A complete rotation takes about 200 million years. The Galaxy's radius is 45,000 light-years, its thickness 3,000 light-years. It bulges in the center. The average distance between two stars is 3 light-years, so this disk is not very dense. The solar system's place is very modest; with the distance from the Sun to Pluto used as a unit of measurement, the Galaxy's radius is 70 million units. The Sun is one of about 200 billion stars in the Galaxy, of which only a few thousand are visible to the naked eye. It is located at two-thirds of the distance from the center to the edge of the disk and it takes part in the overall rotation, the speed of which is 250 km/sec at this radius.

The galactic disk is not uniform. Because of dynamical instabilities, density waves go through it, rather like waves on the surface of water. These are the spiral arms. On the crest of the waves accumulate clouds made of dust, molecules, and hydrogen, where the stars are born. Since we are inside the disk and see it edge-on, the spiral arms are not conspicuous. We had to wait for radio astronomy to find out that we were in a spiral galaxy with four principal arms.

Associations and star clusters — other recognizable structures in the Galaxy — are essentially made up of a small number of young stars, at most several thousands of them; the Pleiades and the association of Orion are the best-known. These systems are born in the spiral arms. Mapping the neigborhood of the Sun allowed us to show that the Sun is located between the exterior arm of Perseus and the major arm of Saggitarius, closer to the latter.

Globular clusters — spherical systems which contain up to a million stars are — on the contrary, very old. They probably formed with, or even before, the Galaxy. They do not rotate with the Galaxy, and their orbits take them well outside the disk. Constellations, unlike clusters and associations, are not homogeneous systems with common properties. Their only purpose is to show a direction in the sky. Thus, the Galaxy's nucleus is located in the constellation of Saggitarius and the cluster of the Pleiades, close to Perseus.

The Milky Way and Andromeda are the two giant spiral galaxies in the Local Group of galaxies, which also includes about twenty smaller galaxies, such as the two Magellanic Clouds. The group of galaxies closest to ours is the Virgo cluster, in the constellation of Virgo, about 30 million light-years away. It contains several thousand galaxies.

The discussion in this book is limited to the search for life in our Galaxy simply because it constitutes, in Kantian terms, an "island universe" — in other words, a rather isolated system. I do not exclude the possibility of extragalactic life, but it is surely easier to start looking for life in our immediate neighborhood.

(Observatoire de Haute-Provence, Centre National de la Recherche Scientifique)

The galaxy NGC 2336. This spiral galaxy, located 48 million light-years from us, resembles our Galaxy: a barred central region and well-developed spiral arms. The bright points around it are stars from our own Galaxy.

Bibliography

General and bibliographic works

Cameron, A.G.W. (editor). *Interstellar communication.* Benjamin, 1963.

Kuiper, T.B.H., and Brin, G.D. "Resource letter ETC-1: Extraterrestrial civilization." *American Journal of Physics* 57(1): 12, 1989.

Mallove, E.P., Connors, M.M., Forward, R.L., and Paprotny, Z. A Bibliography on the Search for Extraterrestrial Intelligence. NASA report RP-1021, 1978.

Mallove, E.P., Forward, R.L., Paprotny, Z., and Lehman, J. "Interstellar travel and communication: A bibliography." *Journal of the British Interplanetary Society* 33: 201, 1980.

G. Marx (editor). *Bioastronomy: The Next Steps.* IAU Colloquium 99, Kluwer, 1988.

Papagiannis, M.D. (editor). The Search for Extraterrestrial Life: Recent Developments. Symposium 112 of the International Astronomical Union, 1985.

Paprotny, Z., Lehman, J., and Prytz, J. "Interstellar travel and communication bibliography — 1984 update." *Journal of the British Interplanetary Society* 37: 502, 1984.

Paprotny, Z., Lehman, J., and Prytz, J. "Interstellar travel and communication bibliography — 1985 update." *Journal of the British Interplanetary Society* 39: 427, 1985.

Ponnamperuma, C., and Cameron, A.G.W. (editors). *Interstellar Communication: Scientific Perspectives.* Houghton Mifflin, 1974.

Rubtsov, V.V., and Ursul, A.D. *Problema vnezemnykh tsivilizatsij.* Shtiintsa, Kishinev (USSR), 1984.

Sagan, C. *The Cosmic Connection*. Doubleday, 1973.

Schatzman, E. *Les enfants d'Uranie*. Seuil, 1986.

Shklovskii, I.S., and Sagan, C. *Intelligent life in the universe*. Holden-Day, 1966.

Project Cyclops. NASA report CR-114445, 1973.

Life in the Universe. NASA report CP-2156, 1981.

L'Astronomie, 103, December 1989. Special issue on exobiology.

Review articles

Papagiannis, M.D. "Commission 51: Search for Extraterrestrial Life." *IAU Transactions* 19A: 713, 1985.

Sagan, C., and Drake, F. "The Search for Extraterrestrial Intelligence." *Scientific American* 232 (5): 80, 1975.

"Life in the universe." In *The Cambridge Encyclopaedia of Astronomy* (Jonathan Cape, 1977).

Chapter 1

Anders, E. "Pre-biotic organic matter from comets and asteroids." *Nature* 342: 255, 1989.

Bonny, P. "Nouvelles collections de micrométéorites groenlandaises et antarctiques." *L'Astronomie* 103: 498, 1989.

Boureau, E. "Au précambrien: naissance du monde vivant." *La Recherche* 68: 541, 1976.

Boureau, E. "Biogenèse." In *Encyclopédia Universalis*, 1980.

Boureau, E. "Les fossiles précambriens et les origines de l'organisation biologique." *La Vie des Sciences, Comptes-Rendus de l'Académie des Sciences, série générale* 1: 431, 1984.

Brownlee, D.E. "Interplanetary dust: Its physical nature and entry into the atmosphere of terrestrial planets." In *Comets and the Origin of Life*, ed. C. Ponnamperuma (Reidel, 1981).

Buhl, D., and Ponnamperuma, C. "Interstellar molecules and the origin of life." *Space Life Sciences* 3: 157, 1971.

Butler, R.N. "On the place of organic life in the universe and the possibility of extraterrestrial life." *Irish Astronomical Journal* 14: 177, 1982.

Delsemme, A.H. "Are comets connected to the origin of life ?" In *Comets and the Origin of Life*, ed. C. Ponnamperuma (Reidel, 1981).

Desbruyères, D. "La vie dans les évents marins." *L'Astronomie* 103: 531, 1989.

Dickerson, R.E. "Chemical evolution and the origin of life." *Scientific American* 239 (3): 70, 1978.

Greenberg, J.M. "Chemical evolution of interstellar dust: a source of prebiotic material ?" In *Comets and the Origin of Life*, ed. C. Ponnamperuma (Reidel, 1981).

Greenberg, J.M., and Weber, P. "Panspermia: A biological approach." In *The Search for Extraterrestrial Life: Recent Developments*, ed. M.D. Papagiannis (IAU, 1985).

Kamminga, H. "Life from space: A history of panspermia." *Vistas in Astronomy* 26: 67, 1982.

Knoll, A.H. "The precambrian evolution of terrestrial life." In *The Search for Extraterrestrial Life: Recent Developments*, ed. M.D. Papagiannis (IAU, 1985).

Lake, J. "Origin of the eukaryotic nucleus determined by rate-invariant analysis of rRNA sequences." *Nature* 331: 184, 1988.

Lawless, J.G., Folsome, C.E., and Kvenvolden, K.N. "Organic matter in meteorites." *Scientific American* 226 (6): 38, 1972.

Mukhin, L.M., Gerasimov, M.V., and Safonova, E.N. "Origin of precursors of organic molecules during evaporation of meteorites and mafic terrestrial rocks." *Nature* 340: 46, 1989.

Parkin, D.W., and Tilles, D. "Influx measurement of extraterrestrial material." *Science* 159: 936, 1968.

Ponnamperuma, C. (editor). *Comets and the Origin of Life*. Reidel, 1981.

Ponnamperuma, C. (editor). *Cosmochemistry and the Origin of Life*. Reidel, 1982.

Ponnamperuma, C. "Cosmochemistry and the origin of life." In *Cosmochemistry and the Origin of Life*, ed. C. Ponnamperuma (Reidel, 1981).

Ponnamperuma, C., and Ochiai, E. "Comets and origin of life." In *Comets*, ed. L.L. Wilkening (University of Arizona Press, 1982).

Roberts, J.A. "Supernovae and life." *Nature* 308: 318, 1984.

Rubenstein, E., Bonner, W.A., Noves, H.P., and Brown G.S. "Supernovae and life." *Nature* 306: 118, 1983.

Schidlowski, M. "A 3,800-million-year isotopic record of life from carbon in sedimentary rocks." *Nature* 333: 313, 1988.

Wallis, M.K., Wickramasinghe, N.C., Hoyle, F., and Rabilizirov, R. "Biologic versus abiotic models of cometary grains." *Monthly Notices of the Royal Astronomical Society* 238: 1165, 1989. (See also scientific correspondence in *Nature* 328: 117, 1987.)

Wolstencroft, R.D. "Astronomical sources of circularly polarized light and their role in determining molecular chirality on earth." In *The Search for Extraterrestrial Life: Recent Developments*, ed. M.D. Papagiannis (IAU, 1985).

"Did life start from clay?" *Newsweek* 105 (15):50, 1985.

Chapter 2

Begley, S. "Science contra Darwin." *Newsweek* 105 (14): 50, 1985.

Eldredge, N. "Progress in evolution ?" *New Scientist* 110 (1511): 54, 1986.

Girardon, J. "Darwin et les extraterrestres." *Sciences et Avenir* 443: 85, 1984.

Halstead, B. "Anti-Darwinian theory in Japan." *Nature* 317: 587, 1985.

Jacob, F. "Evolution and tinkering." *Science* 196: 1161, 1977.

Thuillier, P. "Darwin chez les Samouraï." *La Recherche* 17: 1276, 1986.

Tintant, H. "L'évolution des céphalopodes: Gradualisme ou ponctualisme?" *La vie des Sciences, Comptes-Rendus de l'Académie des Sciences, série générale* 2: 409, 1985.

Chapter 3

Alvarez, L.W. "Mass extinctions caused by large bolide impacts." *Physics Today* 40 (7): 24, 1987.

Alvarez, W., and Muller, R.A. "Evidence from crater ages for periodic impacts the Earth." *Nature* 308: 718, 1984.

Alvarez, L.W., Alvarez, W., Asaro, F., and Michel, H.V. "Extraterrestrial cause for the cretaceous-tertiary extinction." *Science* 208: 1095, 1980.

Audouze, J. "Le géologue, l'astrophysicien et la fin des dinosaures." *La Recherche* 12: 78, 1981.

Cherfas, J. "Extinction patterns of evolution." *New Scientist*, October 3, 1985:48.

Clark, D., Hunt, G., and McCrea, W. "Celestial chaos and terrestrial catastrophes." *New Scientist* 80: 861, 1978.

Clube, S.V.M., and Napier, W.M. "Terrestrial catastrophism: Nemesis or Galaxy ?" *Nature* 311: 635, 1984.

Crawford, A.R. "Spiral arms, comets and terrestrial catastrophism: A Discussion." *Quarterly Journal of the Royal Astronomical Society* 26: 53, 1985.

Crowley, T.J., and North, G.R. "Abrupt climate change and extinction events in earth history." *Science* 240: 996, 1988.

Crutzen, P.J. "Acid rain at the K/T boundary." *Nature* 330: 108, 1987.

Davis, M., Hut, P., and Muller, R.A. "Extinction of species by periodic comet showers." *Nature* 308: 715, 1984.

Delsemme, A.H. "Une explication astronomique à la disparition des dinosaures." *L'Astronomie* 99: 3, 1985.

Donovan, S.K. "Iridium anomalous no longer?" *Nature* 326: 331, 1987.

Fremlin, J. "Dinosaur death: The unconscious factor." *New Scientist* 81: 259, 1979.

Hallam, A. "The causes of mass extinctions." *Nature* 308: 686, 1984.

Halliday, I., Blackwell, A.T., and Griffin, A.A. "Meteorite impacts on humans and on buildings." *Nature* 318: 317, 1985.

Hess, J., Bender, M.L., and Schilling, J.-G. "Evolution of the ratio of strontium-87 to strontium-86 in seawater from cretaceous to present." *Science* 231: 979, 1986.

Hut, P., Alvarez, W., Elder, W.P., Hansen, T., Kauffman, E.G., Keller, G., Shoemaker, E.M., and Weissman, P.R. "Comet showers as a cause of mass extinctions." *Nature* 329: 118, 1987.

Jablonski, D. "Background and mass extinctions: The alternation of macroevolutionary regimes." *Science* 231: 129, 1986.

Kyte, F.T., and Wasson, J.T. "Accretion rate of extraterrestrial matter: Iridium deposited 33 to 67 million years ago." *Science* 232: 1225, 1986.

Lutz, T.M. "The magnetic reversal record is not periodic." *Nature* 317: 404, 1985.

Muller, R.A. "Evidence for a solar companion star." In *The Search for Extraterrestrial Life: Recent Developments*, ed. M.D. Papagiannis (IAU, 1985).

Officer, C.B., and Drake, C.L. "Terminal cretaceous environmental events." *Science* 227: 1161, 1985.

Officer, C.B., Hallam, A., Drake, C.L., and Devine, J.D. "Late Cretaceous and paroxysmal Cretaceous/Tertiary extinctions." *Nature* 326: 143, 1987.

Opik, E.J. "Our cosmic destiny." *Irish Astronomical Journal* 11: 113, 1973.

Poupeau, G. "Une horloge astronomique rythme-t-elle la disparition des espèces?" *La Recherche* 16: 674, 1985.

Rampino, M.R., and Stothers, R.B. "Terrestrial mass extinctions, cometary impacts and the sun's motion perpendicular to the galactic plane." *Nature* 308: 709, 1984.

Rampino, M.R., and Stothers, R.B. "Flood basalt volcanism during the past 250 million years." *Science* 241: 663, 1988.

Raup, D.M. "Magnetic reversals and mass extinctions." *Nature* 314: 341, 1985.

Raup, D.M.,and Sepkoski, J.J.,Jr. "Periodic extinctions of families and genera." *Science* 231: 833, 1986.

Renard, M., and Rocchia, R. "Extinction des espèces au secondaire: la Terre dans un nuage interstellaire?" *La Recherche* 15: 393, 1984.

Rocchia, R., and Boclet, D. "Des dinosaures aux pluies de comètes." *Ciel et Espace* 207: 28, 1986.

Russel, D.A. "The mass extinctions of the late mesozoic." *Scientific American* 48, January 1982.

Schwarz, R.D., and James, P.B. "Periodic mass extinctions and the Sun's oscillation about the galactic plane." *Nature* 308: 712, 1984.

Schwarzschild, B. "Do asteroid impacts trigger geomagnetic reversals?" *Physics Today* 40 (2):17, 1987.

Sepkoski, J.J., Jr. "Some implications of mass extinction for the evolution of complex life." In *The Search for Extraterrestrial Life: Recent Developments*, ed. M.D. Papagiannis (IAU, 1985).

Thaddeus, P., and Chanan, G.A. "Cometary impacts, molecular clouds, and the motion of the Sun perpendicular to the galactic plane." *Nature* 314: 73, 1985.

Torbett, M.V., and Smoluchowski, R. "Orbital stability of the unseen solar companion linked to periodic extinction events." *Nature* 311: 641, 1984.

Tucker, W.H. "Astrophysical crises in the evolution of life in the Galaxy." NASA report CP-2156, 1981.

Van Den Bergh, S. "Life and death in the inner solar system." *Publications of the Astronomical Society of the Pacific* 101: 500, 1989.

Van Valen, L.M. "The case against impact extinction." *Nature* 311: 18, 1984.

Weissman, P.R. "Terrestrial impactors at geological boundary events: Comets or asteroids?" *Nature* 314: 517, 1985.

Whitmire, D.P., and Jackson, A.A.,IV. "Are periodic mass extinctions driven by a distant solar companion?" *Nature* 308: 713, 1984.

Whitmire, D.P., and Matese, J.J. "Periodic comet showers and planet X." *Nature* 313: 36, 1985.

Wolbach, W.S., Gilmour, I., Anders, E., Orth, C.J., and Brooks, R.R. "Global fire at the Cretaceous-Tertiary boundary." *Nature* 334: 665, 1988.

Zhao, M., and Bada, J.L. "Extraterrestrial amino acids in Cretaceous/Tertiary boundary sediments at Stevns Klint, Denmark." *Nature* 339: 463, 1989.

Chapter 4

Balébanov, V. "Mission Phobos: un pas vers Mars." *La Recherche* 19: 396, 1988.

Banks, P.M., and Ride, S.K. "Soviets in space." *Scientific American* 260 (2): 18, 1989.

Barnola, J.M., Raynaud, D., Korotkevich, Y.S., and Lorius, C. "Vostok ice core provides 160,000-year record of atmospheric CO_2." *Nature* 329: 408, 1987.

Berger, A. "L'hiver nucléaire." *La Recherche* 17 (179): 880, 1986.

de Goursac, O. "Objectif 2017: des hommes sur Mars!" *Ciel et Espace* 216: 33, 1987.

Dickinson, R.E., and Cicerone, R.J. "Future global warming from atmospheric trace gases." *Nature* 319: 109, 1986.

Dyson, F. "Human consequences of the exploration of space." *Bulletin of the Atomic Scientists* 25 (7): 8, 1969.

Eshleman, V.R. "Colonization of space: Destiny or folly?" *Nature* 324: 115, 1986.

Finney, B.R. "Exponential expansion: Galactic destiny or technological hubris ?" In *The Search for Extraterrestrial Life: Recent Developments*, ed. M.D. Papagiannis (IAU, 1985).

Guillen, M. "Moon mines, space factories and colony L5." *Science News* 110: 124, 1976.

Heath, D.F. "Non-seasonal changes in total column ozone from satellite observations, 1970–86." *Nature* 332: 219, 1988.

Khalil, M.A.K., and Rasmussen, R.A. "Carbon monoxide in the earth's atmosphere: Indications of a global increase." *Nature* 332: 242, 1988.

Kuo, C., Lindberg, C., and Thomson, D.J. "Coherence established between atmospheric carbon dioxide and global temperature." *Nature* 343: 709, 1990.

Lebeau, A. "L'astronaute et le robot." *Futuribles*, September 1986.

Marshall, E. "Nuclear winter debate heats up." *Science* 235: 271, 1987.

Oberg, J.E. "Un homme sur Mars." *La Recherche* 17: 510, 1986.

O'Neill, G.K. "Colonization of space." *Physics Today* 27 (9): 32, 1974.

O'Neill, G.K. "Space colonies and energy supply to the Earth." *Science* 190: 943, 1975.

Papagiannis, M.D. "Natural selection of stellar civilizations by the limits of growth." *Quarterly Journal of the Royal Astronomical Society* 25: 309, 1984.

Pyle, J.A., and Farman, J.C. "Antarctic chemistry to blame." *Nature* 329: 103, 1987.

Schneider, S.H. "The greenhouse effect: Science and policy." *Science* 243: 771, 1989.

Schneider, S.H., and Thompson, S.L. "Simulating the climatic effects of nuclear war." *Nature* 333: 221, 1988.

Stolarski, R.S. "The antarctic ozone hole." *Scientific American* 258 (1): 20, 1988.

Sullivan,S. "Nature's revenge." *Newsweek* 109 (9): 40, 1987.

Sundquist, E.T. "Ice core links CO_2 to climate." *Nature* 329: 389, 1987.

Turco, R.P., Toon, O.B., Ackerman, T.P., Pollack, J.B.,and Sagan, C. "Nuclear winter: Global consequences of multiple nuclear explosions." *Science* 222: 1283, 1983.

Turco, R.P., Toon, O.B., Ackerman, T.P., Pollack, J.B., and Sagan, C. "Climate and smoke: An appraisal of nuclear winter." *Science* 247: 166, 1990.

Volz, A., and Kley, D. "Evolution of the Montsouris series of ozone measurements made in the nineteenth century." *Nature* 332: 240, 1988.

von Hörner, S. "Population explosion and interstellar expansion." *Journal of the British Interplanetary Society* 28: 691, 1975.

von Hörner, S. "The general limits of space travel." *Science* 137: 18, 1962.

Chapter 5

Cairns-Smith, G. "Signs of life." *New Scientist* 109 (1489): 34, 1986.

Carlotto, M.J. "Digital imagery analysis of unusual Martian surface features." *Applied Optics* 27: 1926, 1988.

Davies, R.E., Delluva, A.M., and Koch, R.H. "No valid evidence exists for interstellar proteins, bacteria, etc." In *The Search for Extraterrestria Life: Recent Developments*, ed. M.D. Papagiannis (IAU, 1985).

Horowitz, N.H. "La vie sur Mars." In *Le système solaire* (Bibliothèque pour la Science).

Reynolds, R.T., Squyres, S.W., Colburn, D.S., and McKay, C.P. "On the habitability of Europa." *Icarus* 56: 246, 1983.

Sagan, C., Thompson, W.R., and Khare, B.N. "Titan's organic chemistry." In *The Search for Extraterrestrial Life: Recent Developments*, ed. M.D. Papagiannis (IAU, 1985).

Schneider, J. "A model for a non-chemical form of life: Crystalline physiology." *Origins of Life* 8: 33, 1977.

Chapter 6

Aumann, H.H. "Prospects of infrared prospecting for planets." *Nature* 328: 208, 1987.

Aumann, H.H. "Protoplanetary material around nearby stars." In *The Search for Extraterrestrial Life: Recent Developments*, ed. M.D. Papagiannis (IAU,1985).

Becklin, E.E., and Zuckerman, B. "A low-temperature companion to a white dwarf star." *Nature* 336: 656, 1988.

Beckwith, S., and Sargent, A. "HL Tauri, a site for planet formation?" *Mercury* 16: 178, 1987.

Black, D.C. "A review of the scientific rationale and methods used in the search for other planetary systems." In *The Search for Extraterrestrial Life: Recent Developments*, ed. M.D. Papagiannis (IAU, 1985).

BMS. "Infrared evidence for protoplanetary rings around seven stars." *Physics Today* 37 (5): 17, 1984.

Burke, B.F. "Detection of planetary systems and the search for evidence of life." *Nature* 322: 340, 1986.

Campbell, B., and Garrison, R.F. "On the inclination of extra-solar planetary orbits." Preprint, Dominion Astrophysical Observatory, Canada, 1984.

Campbell, B., Walker, G.A.H., Pritchet, C., and Long, B. "The search for low-mass stellar companions with the HF precision velocity technique." Workshop on the Astrophysics of Brown Dwarfs, Fairfax, Virginia, October 1985.

Campbell, B., and Walker, G.A.H. "Stellar radial velocities of high precision: Techniques and results." In *Stellar Radial Velocities*, IAU Colloquium 88, October 1984.

Couper, H. "In search of solar systems." *New Scientist* 112 (1534): 34, 1986.

Diner, D.J., and Appleby, J.F. "Prospecting for planets in circumstellar dust: Sifting the evidence from β Pictoris." *Nature* 322: 436, 1986.

Duquennoy, A., and Mayor, M. "Duplicity of solar-like stars in the solar neighborhood." In *Proceedings of the XI European Regional Astronomy Meeting*, ed. M. Vazquez (Cambridge University Press).

Encrenaz, T. "La recherche d'autres systèmes solaires." *La Recherche* 14: 1597, 1983.

Harvey, P.M., Wilking, B.A., and Joy, M. "On the far-infrared excess of Vega." *Nature* 307: 441, 1984.

Latham, D.W., Mazeh, T., Stefanik, R.P., Mayor, M., and Burki, G. "The unseen companion of HD 114762: A probable brown dwarf." *Nature* 339: 38, 1989.

Mc Carthy, D.W., and Henry, T.J. "Direct infrared observations of the very low mass object Gliese 623B." *Astrophysical Journal* 319: L95, 1987.

Nakano, T. "Formation of planets around stars of various masses. I. Formulation and a star of one solar mass." *Monthly Notices of the Royal Astronomical Society* 224: 117, 1987.

Page, T. "Detecting distant planets with space telescope." In *The Search for Extraterrestrial Life: Recent Developments*, ed. M.D. Papagiannis (IAU, 1985).

Perrier, C., and Mariotti, J.-M. "On the binary nature of Van Biesbroek 8." *Astrophysical Journal* 312: L23, 1987.

Reeves, H. "Une planète bien curieuse." *Ciel et Espace* 204: 5, 1985.

Struck-Marcell, C.C. "SiO masers in late-type giant stars: Jovian planets engulfed." *Astrophysical Journal* 330: 986, 1988.

Tayler, R.J. "Les naines brunes." *La Recherche* 19: 326, 1988.

Zuckerman, B., and Becklin, E.E. "Excess infrared radiation from a white dwarf: An orbiting brown dwarf?" *Nature* 300: 138, 1987.

Chapter 7

Freeman, J., and Lampton, M. "Interstellar archeology and the prevalence of intelligence." *Icarus* 25: 368, 1975.

Harrington, R.S., and Harrington, B.J. "Can we find a place to live near a multiple star ?" *Mercury* 7 (6), 1978.

Hart, M.H. "Habitable zones about main sequence stars." *Icarus* 37: 351, 1979.

Hohlfeld, R.G., and Terzian, Y. "Multiple stars and the number of habitable planets in the Galaxy." *Icarus* 30: 598, 1977.

Kastings, J.F., Toon, O.B., and Pollack, J.B. "How climate evolved on the terrestrial planets." *Scientific American* 258 (2):46, 1988.

Kreifeld, J.G. "A formulation for the number of communicative civilizations in the Galaxy." *Icarus* 14: 419, 1971.

Lindley, D. "Is the Earth alive or dead?" *Nature* 332: 483, 1988.

Lovelock, J. "Gaia: The world as a living organism." *New Scientist* 112: 1539, 1986.

Oliver, B.M. "Proximity of galactic civilizations." *Icarus* 25: 360, 1975.

Schneider, S.H., and Thompson, S.L. "Cosmic conclusions from climatic models: Can they be justified?" *Icarus* 41: 456, 1980.

Shklovskij, I.S. "Problema vnezemnykh tsivilizatsei i ee filisofskie aspekti." *Voprosy Filosofii* 2: 76, 1973.

Sturrock, P.A. Uncertainty Estimates of the Number of Extraterrestrial Civilizations. SUIPR report 808, 1980.

Wallenhorst, S.G. "The Drake equation re-examined." *Quarterly Journal of the Royal Astronomical Society* 22: 380, 1981.

Walters, C., Hoover, R.A., and Kotra, R.K. "Interstellar colonization: A new parameter for the Drake equation ?" *Icarus* 41: 193, 1980.

Chapter 8

Ball, J. "The zoo hypothesis." *Icarus* 19: 347, 1973.

Brin, G.D. "The great silence: The controversy concerning extraterrestrial intelligent life." *Quarterly Journal of the Royal Astronomical Society* 24: 283, 1983.

Clarke, J.N. "Extraterrestrial intelligence and galactic nuclear activity." *Icarus* 46: 94, 1981.

Cox, L.J. "An explanation for the absence of extraterrestrials on Earth." *Quarterly Journal of the Royal Astronomical Society* 17: 201, 1976.

Hart, M.H. "An explanation for the absence of extraterrestrials on Earth." *Quarterly Journal of the Royal Astronomical Society* 16: 128, 1975.

Jones, E.M. "Colonization of the Galaxy." *Icarus* 28: 421, 1976.

Jones, E.M. "Discrete calculations of interstellar migration and settlement." *Icarus* 46: 328, 1981.

Newman, W.I., and Sagan, C. "Galactic civilizations: Population dynamics and interstellar diffusion." *Icarus* 46: 293, 1981.

Papagiannis, M.D. "Are we all alone, or could they be in the asteroid belt?" *Quarterly Journal of the Royal Astronomical Society* 19: 277, 1978.

Sagan, C. "Direct contact among galactic civilizations by relativistic interstellar spaceflight." *Planetary and Space Science* 11: 485, 1963.

Sagan, C. "On the detectivity of advanced galactic civilizations." *Icarus* 19: 350, 1973.

Sagan, C., and Newman, W.I. "The solipsist approach to extraterrestrial intelligence." *Quarterly Journal of the Royal Astronomical Society* 24: 113, 1983.

Schwartzman, D. "The absence of extraterrestrials on earth and the prospects for CETI." *Icarus* 32: 473, 1977.

Stephenson, D.G. "Extraterrestrial cultures within the solar system." *Quarterly Journal of the Royal Astronomical Society* 20: 422, 1979.

Tipler, F. "Extraterrestrial intelligent beings do not exist." *Quarterly Journal of the Royal Astronomical Society* 21: 267, 1980.

Tipler, F. "Extraterrestrial intelligent beings do not exist." *Physics Today*, April 1981, p. 9. (See also letters to the editor, March 1982, p. 26.)

Tipler, F. "A brief history of the extraterrestrial intelligence concept." *Quarterly Journal of the Royal Astronomical Society* 22: 133, 1981.

Tipler, F. "Additional remarks on extraterrestrial intelligence." *Quarterly Journal of the Royal Astronomical Society* 22: 279, 1981.

Tipler, F. "The most advanced civilization in the Galaxy is ours." *Mercury* 11 (1): 5, 1982.

Chapter 9

Condon, E.U., et al. *Scientific Study of Unidentified Flying Objects* (Dutton). Reviews: J. A. Hynek, *Bulletin of the Atomic Scientists* 25 (4): 39, 1969; G. Rothberg, *Physics Today* 22 (12): 69, 1969; H.Y. Chiu, *Icarus* 11: 447, 1970.

Condon, E.U. "UFOs I have loved and lost." *Bulletin of the Atomic Scientists* 25 (10): 6, 1969.

"Illusion au pays des lumières." *Sciences et Avenir* 56: 6, 198-.

Esterle, A. "Les phénomènes aérospatiaux non identifiés à l'étude en France." *La Recherche* 10: 761, 1979.

Granger, M., and Oberg, J.E. "La NASA et les chasseurs d'OVNI." *La Recherche* 10: 753, 1979.

Margolis, H. "The UFO phenomenon." *Bulletin of the Atomic Scientists* 23 (6): 40, 1967.

Migouline, V. "Les phénomènes aérospatiaux non identifiés à l'étude en Union Soviétique." *La Recherche* 10: 760, 1979.

Pauwells, L., and Bergier, J. *Le Matin des Magiciens.* Gallimard, 1960.

Poher, C. Etude statistique des rapports d'observation du phénomène OVNI. GEPAN, Toulouse, 1976.

Reeves, H. "Le message des OVNI." *La Recherche* 10: 762, 1979.

Rubtsov, V.V. "Building the scientific paleovisitology." *Ancient Skies* 13 (2), 1986.

Sagan, C., "Unidentified flying objects." *Bulletin of the Atomic Scientists* 23 (6): 43, 1967.

Sturrock, P.A. "Extraterrestrial intelligent life." *Quarterly Journal of the Royal Astronomical Society* 19: 521, 1978.

Temple, R.K.S. *The Sirius Mystery.* (Sidgwick and Jackson, 1976). Review: M. Ovenden, *Nature* 261: 617, 1976.

"On the UFOs." *Bulletin of the Atomic Scientists* 23 (10): 40, 1967.

"Des soucoupes volantes aux OVNI." *Problèmes politiques et sociaux* 450–451: 18, 1982.

Chapter 10

Bates, D.R. "On making radio contact with extraterrestrial civilizations." *Astrophysics and Space Science* 55: 7, 1978.

Bates, D.R. "Difficulty of interstellar radio communication." *Nature* 248: 317, 1974.

Black, D., Tarter, J., Cuzzi, J.N., Conners, M., and Clark, T.A. "Searching for extraterrestrial intelligence: The ultimate exploration." *Mercury* 6 (4): 8, 1977.

Bracewell, R. "Interstellar messengers." *Mercury* 4 (2): 4, 1975.

Cocconi, G., and Morrison, P. "Searching for interstellar communications." *Nature* 184: 844, 1959.

Deardorff, J.W. "Possible extraterrestrial strategy for Earth." *Quarterly Journal of the Royal Astronomical Society* 27: 94, 1986.

Dixon, R.S. "A search strategy for finding extraterrestrial radio beacons." *Icarus* 20: 187, 1973.

Drake, F.D., and Sagan, C. "Interstellar radio communication and the frequency selection problem." *Nature* 245: 257, 1973.

Dyson, F.J. "Search for artificial stellar sources of infrared radiation." *Science* 131: 1667, 1960.

Freitas, R.A., Jr. "Metalaw and interstellar relations." *Mercury* 6 (2): 15, 1977.

Freudenthal, H. "Lincos, design of a language for cosmic intercourse." Review: L. Hogben, *Nature* 192: 826, 1967.

Frisch, D., and Melia, F. "Siblings for SETI." *Icarus* 55: 432, 1983.

Kardashev, N.S. "Transmission of information by extraterrestrial civilizations." *Soviet Astronomy* 8: 217, 1964.

Kuiper, T.B.H., and Morris, M. "Searching for extraterrestrial civilizations." *Science* 196: 616, 1977.

McLaughlin, W.I. "On the timing of an interstellar communication." *Icarus* 32: 464, 1977.

Makovetskii, P.V. "Mutual strategy of search for CETI call signals." *Icarus* 41: 178, 1980.

Makovetskii, P.V. "Structure of call signals of extraterrestrial civilizations." *Soviet Astronomy* 20: 123, 1976.

Makovetskii, P.V. "Nova Cygni 1975: A synchrosignal for extraterrestrial civilizations?" *Soviet Astronomy* 21: 251, 1977.

Murray, B., Gulkis, S., and Edelson, R.E. "Extraterrestrial intelligence: An observational approach." *Science* 199: 485, 1978.

National Astronomy and Ionosphere Center. "The Arecibo message of November, 1974." *Icarus* 26: 462, 1975.

Pace, G.W., and Walker, J.C.G. "Time markers in interstellar communication." *Nature* 254: 400, 1975.

Papagiannis, M.D. "An infrared search in our solar system as part of a more flexible search strategy." In *The Search for Extraterrestrial Life: Recent Developments*, ed. M.D. Papagiannis (IAU, 1985).

Seeger, C.L. "Strategic considerations in SETI, and a microwave approach." *Acta Astronautica* 6: 105, 1979.

Sullivan, W.T., III, and Mighell, K.J. "A Milky Way search strategy for extraterrestrial intelligence." *Icarus* 60: 675, 1984.

USSR Academy of Sciences. "The Soviet CETI program." *Icarus* 26: 377, 1975.

von Hörner, S. "The search for signals from other civilizations." *Science* 134: 1839, 1961.

Chapter 11

Biraud, F. "Bilan du programme SETI à Nançay, 1981–1989." *L'Astronomie* 103: 566, 1989.

Bowyer, S., Zeitlin, G., Tarter, J., Lampton, M., and Welch, W.J. "The Berkeley parasitic SETI program." *Icarus* 53: 147, 1983.

Cohen, N.L., Malkan, M.A., and Dickey, J.M. "A passive SETI in globular clusters at the hydroxyl and water lines." *Icarus* 41: 198, 1980.

Dixon, R.S. "The Ohio SETI program: The first decade." In *The Search for Extraterrestrial Life: Recent Developments*, ed. M.D. Papagiannis (IAU, 1985).

Dixon, R.S., and Cole, D.M. "A modest all-sky search for narrow-band radio radiation near the 21-cm hydrogen line." *Icarus* 30: 267, 1977.

Gray, R.H. "A small SETI radio telescope." *Sky and Telescope*, April 1985: 354.

Heidmann, J. "Emission radio anormale provenant de la direction d'une étoile." *Comptes-Rendus de l'Académie des Sciences*, Série II, 1: 47, 1986.

Heidmann, J. "Le Mega-SETI, une étape majeure en bioastronomie." *Journal des astronomes français* 29: 11, 1987.

Heidmann, J. "Mega-SETI, à la recherche des extraterrestres." *Ciel et Espace* 225: 32, 1988.

Horowitz, P. "A search for ultra-narrowband signals of extraterrestrial origin." *Science* 201: 733, 1978.

Horowitz, P., and Froster, J. "Project Sentinel: An ultra-narrowband SETI at Harvard/Smithsonian." In *The Search for Extraterrestrial Life: Recent Developments*, ed. M.D. Papagiannis (IAU, 1985).

Kent, S., and Uhil, G. Popular Support for the Search for Extraterrestrial Intelligence (SETI). IAF Congress paper IAA-8983.

Papagiannis, M.D. "Recent progress and future plans on the search for extraterrestrial intelligence." *Nature* 318: 135, 1985.

Sweet, W. "Spielberg funds search for ETs by Harvard-Smithsonian." *Physics Today* 39 (1): 80, 1986.

Tarter, J. "SETI observations worldwide." In *The Search for Extraterrestrial Life: Recent Developments*, ed. M.D. Papagiannis (IAU, 1985).

Tarter, J., Cuzzi, J., Black, D., and Clark, T. "A high-sensitivity search for extraterrestrial intelligence at 18 cm." *Icarus* 42: 136, 1980.

Troitskij, V.S., Starodubtsev, A.M., Gershtein, L.I., and Rakhlin, V.L. "Search for monochromatic 927-Mhz radio emission from nearby stars." *Soviet Astronomy* 15: 508, 1971.

Troitskij, V.S., Bondar, L.N., and Starodubtsev, A.M. "The search for sporadic radio emission from space." *Soviet Physics Uspekhi* 17: 607, 1975.

Verschuur, G.L. "A search for narrow band 21-cm wavelength signals from ten nearby stars." *Icarus* 19: 329, 1973.

Chapter 12

Bainbridge, W.S. "Attitudes toward interstellar communication: An empirical study." *Journal of the British Interplanetary Society* 36: 298, 1983.

Hoyle, F.,and Wickramasinghe, N.C. "The case for life as a cosmic phenomenon." *Nature* 322: 509, 1986.

Hoyle, F., Wickramasinghe, N.C., and Al-Mufti, S. "The case for interstellar microorganisms." *Astrophysics and Space Science* 110: 401, 1985.

Maddox, J. "When reference means deference." *Nature* 321: 723, 1986.

Sagan, C. "Extraterrestrial intelligence: An international petition." *Science* 218: 426, 1982.

Sagan, C. "SETI petition." *Science* 220: 462, 1983.

Tarter, J.C. "SETI program." *Science* 220: 359, 1983.

Tipler, F.J. "Extraterrestrial intelligence: A skeptical view of radio searches." *Science* 219: 110, 1983.

Chapter 13

Carr, B.J. "On the origin and purpose of the physical universe." *Irish Astronomical Journal* 15: 237, 1983.

Carr, B.J., and Rees, M.J. "The anthropic principle and the structure of the physical world." *Nature* 278: 605, 1979.

Finney, B. "The impact of contact." International Astronautical Congress, Innsbrück, 1986.

Gale, G. "The anthropic principle." *Scientific American* 245 (6): 154, 1981.

Rees, M. "The anthropic universe." *New Scientist* 115 (1572): 44, 1987.